THREATS FROM SPACE: A REVIEW OF U.S. GOVERNMENT EFFORTS TO TRACK AND MITIGATE ASTEROIDS AND METEORS (PART I & PART II)

HEARING
BEFORE THE

COMMITTEE ON SCIENCE, SPACE, AND TECHNOLOGY
HOUSE OF REPRESENTATIVES
ONE HUNDRED THIRTEENTH CONGRESS

FIRST SESSION

TUESDAY, MARCH 19, 2013
and
WEDNESDAY, APRIL 10, 2013

Serial No. 113–14
and
Serial No. 113–17

Printed for the use of the Committee on Science, Space, and Technology

Available via the World Wide Web: http://science.house.gov

U.S. GOVERNMENT PRINTING OFFICE

80–552PDF   WASHINGTON : 2013

For sale by the Superintendent of Documents, U.S. Government Printing Office
Internet: bookstore.gpo.gov   Phone: toll free (866) 512–1800; DC area (202) 512–1800
Fax: (202) 512–2104   Mail: Stop IDCC, Washington, DC 20402–0001

COMMITTEE ON SCIENCE, SPACE, AND TECHNOLOGY

HON. LAMAR S. SMITH, Texas, *Chair*

DANA ROHRABACHER, California
RALPH M. HALL, Texas
F. JAMES SENSENBRENNER, JR., Wisconsin
FRANK D. LUCAS, Oklahoma
RANDY NEUGEBAUER, Texas
MICHAEL T. McCAUL, Texas
PAUL C. BROUN, Georgia
STEVEN M. PALAZZO, Mississippi
MO BROOKS, Alabama
RANDY HULTGREN, Illinois
LARRY BUCSHON, Indiana
STEVE STOCKMAN, Texas
BILL POSEY, Florida
CYNTHIA LUMMIS, Wyoming
DAVID SCHWEIKERT, Arizona
Thomas Massie, Kentucky
KEVIN CRAMER, North Dakota
JIM BRIDENSTINE, Oklahoma
RANDY WEBER, Texas
CHRIS STEWART, Utah
VACANCY

EDDIE BERNICE JOHNSON, Texas
ZOE LOFGREN, California
DANIEL LIPINSKI, Illinois
DONNA F. EDWARDS, Maryland
FREDERICA S. WILSON, Florida
SUZANNE BONAMICI, Oregon
ERIC SWALWELL, California
DAN MAFFEI, New York
ALAN GRAYSON, Florida
JOSEPH KENNEDY III, Massachusetts
SCOTT PETERS, California
DEREK KILMER, Washington
AMI BERA, California
ELIZABETH ESTY, Connecticut
MARC VEASEY, Texas
JULIA BROWNLEY, California
MARK TAKANO, California
VACANCY

CONTENTS

Tuesday, March 19, 2013

	Page
Witness List	2
Hearing Charter	3

Opening Statements

Statement by Representative Lamar S. Smith, Chairman, Committee on Science, Space, and Technology, U.S. House of Representatives 5
    Written Statement 6
Statement by Representative Eddie Bernice Johnson, Ranking Member, Committee on Science, Space, and Technology, U.S. House of Representatives 6
    Written Statement 7
Statement by Representative Donna F. Edwards, Committee on Science, Space and Technology, U.S. House of Representatives 8
    Written Statement 8

Witnesses:

The Honorable John P. Holdren, Director, Office of Science and Technology Policy, Executive Office of the President
    Oral Statement 10
    Written Statement 12

Gen. William L. Shelton, Commander, U.S. Air Force Space Command
    Oral Statement 19
    Written Statement 20

The Honorable Charles F. Bolden, Jr., Administrator, National Aeronautics and Space Administration
    Oral Statement 27
    Written Statement 30

Discussion 38

Appendix I: Answers to Post-Hearing Questions

The Honorable John P. Holdren, Director, Office of Science and Technology Policy, Executive Office of the President 64
Gen. William L. Shelton, Commander, U.S. Air Force Space Command 73
The Honorable Charles F. Bolden, Jr., Administrator, National Aeronautics and Space Administration 82

Appendix II: Additional Material for the Record

Submitted statement by Representative Steve Stockman, Committee on Science, Space and Technology 94
Letter submitted by Dr. Dante Lauretta, Department of Planetary Sciences, Lunar and Planetary Laboratory 95
Additional responses submitted by The Honorable Charles F. Bolden, Jr., Administrator, National Aeronautics and Space Administration 97

CONTENTS

Wednesday, April 10, 2013

	Page
Witness List	102
Hearing Charter	103

Opening Statements

Statement by Representative Lamar S. Smith, Chairman, Committee on Science, Space, and Technology, U.S. House of Representatives 105
    Written Statement 105
Statement by Representative Eddie Bernice Johnson, Ranking Member, Committee on Science, Space, and Technology, U.S. House of Representatives 106
    Written Statement 107

Witnesses:

Dr. Ed Lu, Chairman and CEO, B612 Foundation
    Oral Statement 108
    Written Statement 112

Dr. Donald K. Yeomans, Manager, Near-Earth Objects Program Office, Jet Propulsion Laboratory
    Oral Statement 117
    Written Statement 119

Dr. Michael F. A'Hearn, Vice-Chair, Committee to Review Near-Earth Object Surveys and Hazard Mitigation Strategies, National Resource Council
    Oral Statement 126
    Written Statement 128

Discussion 136

Appendix I: Answers to Post-Hearing Questions

Dr. Ed Lu, Chairman and CEO, B612 Foundation 150
Dr. Donald K. Yeomans, Manager, Near-Earth Objects Program Office, Jet Propulsion Laboratory 156
Dr. Michael F. A'Hearn, Vice-Chair, Committee to Review Near-Earth Object Surveys and Hazard Mitigation Strategies, National Resource Council 169

Appendix II: Additional Material for the Record

Submitted statement by Representative Steve Stockman, Committee on Science, Space and Technology, U.S. House of Representatives 184
Submitted statement by Representative Donna F. Edwards, Committee on Science, Space and Technology, U.S. House of Representatives 186
Planetary Society Report submitted by Representative Dana Rohrabacher, Committee on Science, Space, and Technology, U.S. House of Representatives 187

THREATS FROM SPACE: A REVIEW OF U.S. GOVERNMENT EFFORTS TO TRACK AND MITIGATE ASTEROIDS AND METEORS, PART I

TUESDAY, MARCH 19, 2013

House of Representatives,
Committee on Science, Space, and Technology,
Washington, D.C.

The Committee met, pursuant to call, at 10:11 a.m., in Room 2318 of the Rayburn House Office Building, Hon. Lamar Smith [Chairman of the Committee] presiding.

LAMAR S. SMITH, Texas
CHAIRMAN

EDDIE BERNICE JOHNSON, Texas
RANKING MEMBER

Congress of the United States
House of Representatives
COMMITTEE ON SCIENCE, SPACE, AND TECHNOLOGY
2321 RAYBURN HOUSE OFFICE BUILDING
WASHINGTON, DC 20515-6301
(202) 225-6371
www.science.house.gov

U.S. House of Representatives
Committee on Science, Space, and Technology

Threats from Space:
A Review of U.S. Government Efforts
to Track and Mitigate Asteroids and Meteors, Part 1

Tuesday, March 19, 2013
10:00 a.m. to 12:00 p.m.
2318 Rayburn House Office Building

Witnesses

The Honorable John P. Holdren, Director, Office of Science and Technology Policy, Executive Office of the President

Gen. William L. Shelton, Commander, U.S. Air Force Space Command

The Honorable Charles F. Bolden, Jr., Administrator, National Aeronautics and Space Administration

U.S. House of Representatives
Committee on Science, Space, and Technology

HEARING CHARTER

Threats from Space:
A Review of U.S. Government Efforts
to Track and Mitigate Asteroids and Meteors, Part 1

Tuesday, March 19, 2013
10:00 a.m. – 12:00 p.m.
2318 Rayburn House Office Building

Purpose

At 10:00 am on March 19, 2013, the Committee on Science, Space, and Technology will hold a hearing titled "Threats from Space: A Review of U.S. Government Efforts to Track and Mitigate Asteroids and Meteors, Part 1." This is the first in a series of hearings examining the tracking, characterization and mitigation of Near Earth Objects. The hearing will provide Members of the Committee the opportunity to receive testimony regarding the ongoing work, planned efforts, and coordination procedures within the National Aeronautics and Space Administration, the Office of Science and Technology Policy, and the U.S. Air Force Space Command.

Witnesses:

- **The Honorable John P. Holdren**, Director, Office of Science and Technology Policy, Executive Office of the President

- **Gen. William L. Shelton**, Commander, U.S. Air Force Space Command

- **The Honorable Charles F. Bolden, Jr.**, Administrator, National Aeronautics and Space Administration

Overview

On Friday, February 15, 2013, two events occurred that received worldwide attention. An unforeseen meteor (estimated 50 feet in diameter) exploded in the sky above the Russian city of Chelyabinsk releasing the equivalent of a 300 kiloton bomb, about twenty times the explosive energy of the atomic blast used over the city of Hiroshima. This blast injured nearly 1,200 people and resulted in an estimated $33 million in property damage. On the same day, a small asteroid (150 feet in diameter) discovered by amateur astronomers and tracked closely by NASA passed safely by the Earth, but within the orbital belt of geostationary satellites. Until it entered our atmosphere, the Russian meteor went completely undetected. According to NASA, the two events were unrelated, but raised public awareness of the potential threat from Near Earth Objects (NEOs). Today's hearing will cover the U.S. government's plans and programs to track, classify, and mitigate the threat of NEOs. A second hearing is planned this month to address international, commercial private sector, and philanthropic initiatives to survey the sky for asteroids and comets.

From these two incidents, many questions arose, among them:
- Do we have the tools and technology necessary to detect and track Near Earth Objects?
- How often do we currently observe large meteors entering the atmosphere safely over the ocean?
- Are we tracking the right size objects, specifically the ones that can cause significant harm on Earth?
- Once we identify an object, what are our means of tracking it?
- What are our contingencies and mitigation capabilities if we determine there is a threat to the Earth from a NEO impact?
- What process exists amongst government agencies, both foreign and domestic, in such an instance?

The Science, Space, and Technology Committee has been on the forefront of the issues surrounding Near Earth Objects. For example, the NASA Authorization Acts of 2000 and 2005 directed NASA to conduct a survey of the population of NEOs and study mitigation plans. Astronomers estimate 20,000 potentially hazardous asteroids orbit within the vicinity of the Earth.

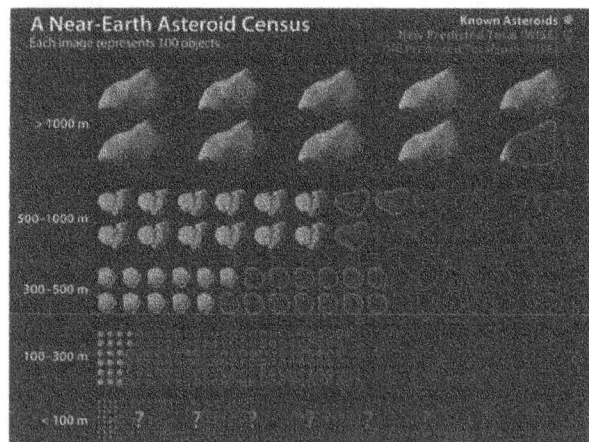

NASA NEO Asteroid Size Model
Credit: NASA/JPL-Caltech
This chart illustrates how infrared is used to more accurately determine an asteroid's size

Chairman SMITH. The Committee on Science, Space, and Technology will come to order. Good morning. I am going to recognize myself for an opening statement, then the Ranking Member, the gentlewoman from Texas, will be recognized as well.

Today's hearing is on a subject important to our Nation and to our world. This is the first hearing of two on space threats to Earth, reviewing U.S. Government efforts to track incoming asteroids and meteors.

Although many may be only aware of this subject due to recent events, it is actually one as old as our planet. And I am going to hold up a copy of Time magazine from nearly 20 years ago where this topic was featured on the cover. Here is Time, ''Cosmic Crash.'' This is 20 years ago. I don't know if they were ahead of their time or not, but in any case, the subject has been around for a while. This was actually given to me by a former staff member, who I had research the subject 20 years ago as well.

Though the issue has been around for a number of years, there are many questions still to be asked and answered. The range of questions are broad and complex, from how to track an object millions of miles away to how to respond if an asteroid or meteor is headed toward Earth.

The two events of Friday, February 15, the harmless flyby of asteroid 2012 DA14 and the not-so-harmless impact of a meteor in Russia, are a stark reminder of the need to invest in space science. The asteroid passed just 17,000 miles from Earth, a distance less than the Earth's circumference. Fifty years ago, we would have had no way of seeing the asteroid coming, and even so, it was discovered by amateur astronomers. The United States has come a long way in its ability to track and characterize asteroids, meteors, comets and meteorites. But we still have a long way to go.

NASA believes it has discovered 93 percent of the largest asteroids in near-Earth orbit, those 1 kilometer or larger, but what about the other seven percent remaining, about 70, or even those smaller than 1 kilometer, estimated to be in the thousands? An asteroid as small as 100 meters could destroy an entire city upon a direct hit. Are we tracking those? The meteor that struck Russia was estimated to be 17 meters, and wasn't tracked at all. The smaller they are, the harder they are to spot, and yet they can be life threatening.

The broad scope of our efforts include participation of governments, research institutions, industries and amateur astronomers in their backyard or on home computers. Some space challenges require innovation, commitment and diligence. This is one of them. And this Committee will strive to continue to lead in this area. For all of the attention and publicity the two events of February 15 received, it was still too late for us to have acted to change the course of the incoming objects. We are in the same position today and for the foreseeable future unless we take actions now that improve our means of detection.

Part of our discussion today is about how to achieve this in the current budget environment. I do not believe that NASA is going to somehow defy budget gravity and get an increase when everyone else is getting cuts. But we need to find ways to prioritize NASA's projects and squeeze as much productivity as we can out of the funds we have. Examining and exploring ways to protect the Earth from asteroids and meteors is a priority for the American people and should be a priority for NASA.

We were fortunate that the events of last month were simply an interesting coincidence rather than a catastrophe. However, we still need to make investments and improvements in our capability to anticipate what may occur decades from now, or tomorrow.

[The prepared statement of Mr. Smith follows:]

PREPARED STATEMENT OF LAMAR S. SMITH, CHAIRMAN, HOUSE COMMITTEE ON SCIENCE, SPACE, AND TECHNOLOGY

Good morning. Today's hearing is on a subject important to our nation and to our world. This is the first hearing of two on Space Threats to Earth, reviewing U.S. Government efforts to track incoming asteroids and meteors.

Although many may be only aware of this subject due to recent events, it is actually one as old as our planet. This is a copy of *TIME Magazine* from nearly 20 years ago (1994) where this topic was featured on the cover.

Though the issue has been around for a number of years, there are many questions still to be asked and answered.

The range of questions are broad and complex, from how to track an object millions of miles away to how to respond if an asteroid or meteor is headed toward Earth.

The two events of Friday, February 15—the harmless flyby of asteroid 2012 DA14 and the not so harmless impact of a meteor in Russia—are a stark reminder of the need to invest in space science.

The asteroid passed just 17,000 miles from Earth, a distance less than the Earth's circumference. Fifty years ago, we would have had no way of seeing the asteroid coming, and even so it was discovered by amateur astronomers.

The U.S. has come a long way in its ability to track and characterize asteroids, meteors, comets and meteorites. But we still have a long way to go. NASA believes it has discovered 93 percent of the largest asteroids in near-Earth orbit, those one kilometer or larger.

But what about the other seven percent remaining, about 70, or even those smaller than one kilometer, estimated to be in the thousands? An asteroid as small as 100 meters could destroy an entire city upon a direct hit. Are we tracking those? The meteor that struck Russia was estimated to be 17 meters, and wasn't tracked at all. The smaller they are, the harder they are to spot, and yet they can be life-threatening.

The broad scope of our efforts include participation of governments, research institutions, industries and amateur astronomers in their backyard or on home computers.

Some space challenges require innovation, commitment and diligence. This is one of them. And this Committee will strive to continue to lead in this area.

For all of the attention and publicity the two events of February 15 received, it was still too late for us to have acted to change the course of the incoming objects. We are in the same position today and for the foreseeable future unless we take actions now that improve our means of detection.

Part of our discussion today is about how to achieve this in the current budget environment.

I do not believe that NASA is going to somehow defy budget gravity and get an increase when everyone else is getting cuts. But we need to find ways to prioritize NASA's projects and squeeze as much productivity as we can out of the funds we have.

Examining and exploring ways to protect the Earth from asteroids and meteors is a priority for the American people and should be a priority for NASA.

We were fortunate that the events of last month were simply an interesting coincidence rather than a catastrophe.

However, we still need to make investments and improvements in our capability to anticipate what may occur decades from now, or tomorrow.

Chairman SMITH. That concludes my opening statement, and the gentlewoman from Texas, Ms. Johnson, is recognized for hers.

Ms. JOHNSON. Thank you very much, Mr. Chairman, and good morning. I would like to welcome each of our witnesses to today's hearing, and I would like to thank you for your patience as we postponed this hearing a couple weeks ago.

As the chairman has indicated, this hearing was called in response to recent events in which a large meteor unexpectedly exploded in the sky over Russia, damaging property and injuring people at almost the same time that a small asteroid passed less than 18,000 miles from Earth's surface. While scientists indicate that those two events apparently were unrelated, they both serve as evidence that we live in an active solar system with potentially hazardous objects passing through our neighborhoods with surprising frequency.

Indeed, there is increasing scientific evidence that impacts by large asteroids and comets have had profound consequences for life on Earth at various times in the past, even contributing to mass extinctions. While such events are very rare, they obviously can cause untold damage, and are not something we want to have happen if we can avoid it.

I think it is our increased scientific understanding of near-Earth objects and their potential to impact the Earth that has led Congress to take this subject seriously in recent years. In that regard, this Committee has taken a leadership role on these issues dating back to the efforts of former Chairman George Brown, Jr. in the early 1990s, a time when references to killer asteroids could still lead to giggles and eye-rolling. Since then, Members on both sides of the aisle, including Representative Rohrabacher, former Chairman Hall and former Representative Giffords have taken an active and productive interest in this topic, and progress has been made.

I hope that today's hearing will provide us with a good update on the Federal Government's efforts to detect, monitor and potentially mitigate such hazardous near-Earth objects. Much has been accomplished over the last decade, and I look forward to hearing about those efforts. In addition, I would like to know if there are additional steps that we should be taking as a country, whether an expanded detection program or international collaborations or other such measures.

Well, we have much to discuss today and a distinguished panel of witnesses to help us in our oversight. I look forward to hearing from each of you.

[The prepared statement of Ms. Johnson follows:]

PREPARED STATEMENT OF RANKING MEMBER EDDIE BERNICE JOHNSON

Good morning. I would like to welcome each of our witnesses to today's hearing. And I would like to thank you for your patience when we were forced to reschedule this hearing in the wake of the Washington snow event two weeks ago.

As the Chairman has indicated, this hearing was called in response to recent events in which a large meteor unexpectedly exploded in the sky over Russia, damaging property and injuring people at almost the same time that a small asteroid passed less than 18,000 miles from Earth's surface. While scientists indicate that those two events apparently were unrelated, they both serve as evidence that we live in an active solar system, with potentially hazardous objects passing through our neighborhood with surprising frequency.

Indeed, there is increasing scientific evidence that impacts by large asteroids and comets have had profound consequences for life on Earth at various times in the past, even contributing to mass extinctions. While such events are very rare, they obviously can cause untold damage, and are not something we want to have happen if we can avoid it.

I think it is our increased scientific understanding of Near Earth Objects and their potential to impact the Earth that has led Congress to take this subject seriously in recent years. In that regard, this Committee has taken a leadership role on these issues dating back to the efforts of former Chairman George Brown, Jr. in the early 1990s—a time when references to "killer asteroids" could still lead to giggles and eye-rolling. Since then, Members on both sides of the aisle, including Rep. Rohrabacher, former Chairman Hall, and former Rep. Giffords have all taken an active and productive interest in this topic, and progress has been made.

I hope that today's hearing will provide us with a good update on the federal government's efforts to detect, monitor, and potentially mitigate such hazardous Near Earth Objects. Much has been accomplished over the last decade, and I look forward to hearing about those efforts.

In addition, I would like to know if there are additional steps that we should be taking as a country, whether an expanded detection program or international collaborations or other such measures.

Well, we have much to discuss today and a distinguished panel of witnesses to help us in our oversight. I look forward to hearing from each of you.

Ms. JOHNSON. At this point I would like to yield the remaining part of my time to Ms. Edwards, the Ranking Member of the Space Subcommittee, for her comments.

Ms. EDWARDS. Thank you, Madam Chairwoman, and thank you, Mr. Chairman.

I just wanted to note for the record, Madam Chairwoman, that this hearing is part one of the Committee's examination of activities related to near-Earth objects. Subcommittee Chairman Palazzo and I will hold a hearing of part two in early April, and so this will be a continuation. And I wanted to note for the record, Madam Chairwoman, that just a month ago after the events that made the news, my colleague, Rush Holt, who is a physicist here in Congress and former Assistant Director of the Princeton Plasma Physics Laboratory, and I coauthored an op-ed that appeared in the Washington Post on February 15 trying to put into plain language what the challenges are, the research challenges, what the threats are so that the American people have some understanding that as both the ranking member and the chairman have noted is not new for this Committee but poses challenges for the American people, especially when it comes to resources.

I think it is very fitting that this Committee is considering U.S. government agency roles and responsibilities in near-Earth object detection, tracking and mitigation, not only because of the recent events, but because we have been at the forefront in setting the U.S. policy on near-Earth objects for the past two decades, and it was this Committee that formulated the provisions in 2008, NASA authorization and subsequent policy direction that called for the Office of Science and Technology Policy to develop policies on emergency response and to recommend a lead agency for protecting the United States, and this depended on NASA, who we always seem to call for 911 assistance in all space matters is in stark contrast to the across-the-board cuts that NASA programs now face under law.

And so Mr. Chairman, I am struck by how this complex planetary protection issue is and how much farther we need to go, and I am looking forward to today's testimony, and with that I yield.

[The prepared statement of Ms. Edwards follows:]

PREPARED STATEMENT OF REPRESENTATIVE DONNA F. EDWARDS

Thank you, Ranking Member Johnson.

It should be noted that this hearing is Part 1 of the Committee's examination of activities related to near-Earth objects (NEOs). Subcommittee Chairman Palazzo and I will hold Part 2 in early April.

It is fitting that this Committee is considering U.S. government agency roles and responsibilities in NEO detection, tracking, and mitigation, not only because of the recent events, but because this Committee has been at the forefront in setting the U.S. policy on NEOs for the past two decades.

The Committee's focus, beginning in the 1990s, has led to NASA's establishment of a system for detection and tracking of large NEOs, such as the 2012 DA14 asteroid. And it was this Committee that formulated the provisions in the 2008 NASA Authorization that called for the Office of Science and Technology Policy to develop policies on emergency response and to recommend a lead agency (or agencies) for protecting the United States from a NEO that is expected to collide with Earth and, if necessary, for implementing a deflection campaign, in consultation with international bodies.

As we will hear today from Dr. Holdren, NASA has a key role.

That should not come as a surprise. NASA's combined scientific, technical, and engineering capability is absolutely essential to informing critical decisions on mitigation of a potentially hazardous object. This dependence on NASA, who we always seem to call for 911 assistance in all space matters, is in stark contrast to the across-the-board sequester cuts to NASA's programs that are now law.

Mr. Chairman, I am struck with how complex this planetary protection issue is and how much farther we need to go. That is why Congress needs to ensure continued investment in and attention to efforts that will address the potential threats of near-Earth objects.

I look forward to hearing from our distinguished group of panelists on the priorities for Congress going forward.

Chairman SMITH. Thank you, Ms. Johnson. Thanks, Ms. Edwards.

Without objection, other Members' opening statements will be made a part of the record.

Our first witness is the Hon. John P. Holdren. Dr. Holdren serves as the Director of the Office of Science and Technology Policy, the Assistant to the President for Science and Technology, and Co-Chair of the President's Council of Advisors on Science and Technology. Prior to his current appointment, he was a professor in both the Kennedy School of Government and the Department of Earth Science at Harvard. Dr. Holdren graduated from M.I.T. with degrees in aerospace engineering and theoretical plasma physics.

General William L. Shelton is the Commander of the United States Air Force Space Command. Prior to assuming his current position, General Shelton was the Assistance Vice Chief of Staff and the Director of the Air Staff at the Pentagon. He currently organizes, equips, trains and maintains mission-ready space and cyberspace forces and capabilities for the North American Aerospace Defense Command and U.S. Strategic Command. General Shelton graduated from the U.S. Air Force Academy with a bachelor's degree in astronautical engineering. He also holds a master's degree in this field from the U.S. Air Force Institute of Technology.

Our final witness is the Hon. Charles F. Bolden, Jr., the Administrator of the National Aeronautics and Space Administration. Administrator Bolden served as a pilot in the Marine Corps, eventually earning the rank of General. In the course of his military career, he participated in several international campaigns. He also tested a variety of ground-attack aircraft until his selection as an astronaut candidate in 1980. Administrator Bolden held a number of positions at NASA. He was able to participate in and support several space shuttle flights, and he traveled to orbit four times aboard the Space Shuttle, twice as a mission commander. For his many achievements, Administrator Bolden was inducted into the U.S. Astronaut Hall of Fame in May of 2006. He earned a bachelor's degree in electrical science from the U.S. Naval Academy and a master's degree in systems management from the University of Southern California.

We welcome you all. Thank you for being here. And Director Holdren, if you will begin?

TESTIMONY OF THE HON. JOHN P. HOLDREN, DIRECTOR, OFFICE OF SCIENCE AND TECHNOLOGY POLICY, EXECUTIVE OFFICE OF THE PRESIDENT

Dr. HOLDREN. Chairman Smith, Ranking Member Johnson, Members of the Committee, I am pleased to be here today to discuss U.S. activities to detect, to track, to characterize near-Earth objects, or NEOs, and to develop the capability to deflect any of dangerous size that are discovered to be on a collision course with the Earth. This is, of course, a particularly timely topic for reasons that all of you mentioned in your opening statements.

Near-Earth objects are defined as those whose orbits bring them within about 31 million miles of the Earth, a third of the distance to the sun, some of them traveling close enough to make an eventual collision a possibility. Those with maximum physical dimension of more than a meter are generally referred to as either asteroids or comets, while smaller objects are referred to as meteoroids. All are called meteors upon fiery transit of the Earth's atmosphere, and the pieces that strike the surface are called meteorites.

Dozens of asteroids a meter or more in size enter the Earth's atmosphere each year, of which only one on the average is as big as 4 meters. Asteroids of these sizes burn up harmlessly high in the atmosphere. Damage on Earth's surface is likely only when the kinetic energy of the object is in the range of a few hundreds of kilotons of TNT equivalent or above. That corresponds at typical closing velocities to a stony asteroid about 15 meters in equivalent diameter.

The 17-meter asteroid that blew up over Russia on February 15 released about 440 kilotons of energy. Asteroids with that much energy strike the Earth only every 100 years or so. Larger events like the 1908 asteroid explosion over Siberia, which released about 15 megatons of energy and leveled trees over an area of more than 850 square miles, are believed to be once-in-a-thousand-years events. If an asteroid explosion of that size were to occur over an urban area, it could cause hundreds of thousands of casualties, but the probability of this occurring is much smaller than the one-in-a-thousand-years probability I just mentioned for one hitting the Earth at all, and that is because land covers only 30 percent of the area of the Earth and urbanized areas cover only two to three percent of the land area.

As a result, the odds of a near-Earth object strike causing massive casualties and destruction of infrastructure are very small, but the potential consequences of such an event are so large that it makes sense to take the risk seriously. Both the Congress and recent Administrations have done so.

In 1998, Congress tasked NASA with locating within 10 years at least 90 percent of all NEOs with a diameter of 1 kilometer or greater, those with the potential to threaten civilization, and in 2005, Congress directed NASA to detect, track, catalog and characterize 90 percent of all NEOs with a diameter of 140 meters or greater by 2020. The 1-kilometer goal was achieved in 2011. The task of detecting 90 percent of NEOs larger than 140 meters is much more challenging but work on it is proceeding apace.

More recent legislation directed the Office of Science and Technology Policy to develop a policy for notifying relevant authorities of an impending threat, to recommend a Federal entity responsible for protecting the Nation from an expected NEO collision, and to implement a policy of threat notification. In an October 2010 letter to this Committee, I reported on our progress on those tasks.

The budget for NASA's Near-Earth Object Observation program has actually increased about fivefold since 2009 from a little less than $4 million to $20.5 million in Fiscal Year 2012. Beyond detection and tracking of potentially threatening objects, moreover, the Administration is committed to exploring and developing the capabilities necessary to protect the Earth in general and the United States in particular from NEO threats. NASA coordinates this work with the Departments of Defense, State and Homeland Security including the latter's Federal Emergency Management Agency.

I thank the Committee for its continued support and its interest in this issue, and I will be pleased to take any questions that the Members may have.

[The prepared statement of Dr. Holdren follows:]

Statement of Dr. John P. Holdren
Director, Office of Science and Technology Policy
Executive Office of the President of the United States
to the
Committee on Science, Space, and Technology
United States House of Representatives
on
March 19, 2013
(updated)

Chairman Smith, Ranking Member Johnson, and Members of the Committee, I am pleased to be here with you today to discuss the status and coordination of U.S. activities to detect, track, and characterize near-Earth objects (NEO) and to develop the capability to deflect any of dangerous size discovered to be on a collision course with Earth. This is a very timely topic — as underscored by the asteroid explosion over Russia on February 15 and the close flyby of an even larger asteroid the same day — and I am looking forward to sharing with you the Administration's perspective on this issue.

I want to start by acknowledging the emphasis that the Congress has placed on understanding and mitigating the threat from NEOs. I thank you for working with the Administration to address this important topic. Through multiple pieces of legislation, Congress has provided direction to pursue enhanced NEO-detection activities and assign responsibility for threat mitigation.

The NASA Authorization Act of 2005 (in a section labeled the George E. Brown, Jr. Near-Earth Object Survey Act) directed NASA to detect, track, catalogue, and characterize 90 percent of all NEOs with a diameter of 140 meters or greater by 2020. This legislation extends Congressional direction from 1998 that tasked NASA with locating at least 90 percent of all NEOs with a diameter of one kilometer or greater — those judged by many experts to have the potential to threaten civilization — within ten years.

The one-kilometer goal was achieved in 2011, with statistical calculations indicating that more than 90 percent of near-Earth objects of this size had been found. The task of detecting 90 percent of NEOs larger than 140 meters is much more challenging, and I will describe the United States' efforts on this front later in this testimony.

More-recent legislation has focused on government-wide preparations to address the threat of a NEO impact. The NASA Authorization Act of 2008 directed the Office of Science and Technology Policy (OSTP) to develop a policy for notifying Federal agencies and relevant emergency-response institutions of an impending threat; it also tasked OSTP with recommending a Federal agency or agencies to be responsible for protecting the United States from an expected NEO collision and implementing a deflection campaign, should one be necessary.

Building on the 2008 language, the NASA Authorization Act of 2010 called for OSTP to implement the policy on threat notification and assign a Federal agency or agencies to be responsible for protecting the United States in the event of a potential collision. In an October 2010 letter to this Committee, I reported that OSTP, in concert with the National Security Staff

(NSS) and the Office of Management and Budget (OMB), had developed an approach I'll outline later in this testimony for delegating responsibility and notifying Federal agencies and relevant emergency-response institutions of an impending NEO threat. In addition, the President's 2010 National Space Policy reinforced NASA's role to "pursue capabilities, in cooperation with other departments, agencies, and commercial partners, to detect, track, catalog, and characterize" NEOs.

Several Federal departments and agencies have significant roles in the pursuit of these goals and they cooperate in important ways. NASA sponsors various activities relating to the search for NEOs, including the collection and correlation of NEO orbit data, precision tracking and characterization of NEOs, and assessments of NEO orbits and impact probabilities in conjunction with other U.S. government agencies including the U.S. Air Force and other Department of Defense (DoD) organizations, the U.S. Department of Energy (DOE), and the National Science Foundation (NSF), each of which plays a key role in funding ground-based astronomical assets that are used to detect and track NEOs.

BACKGROUND

Near-earth objects (NEOs) are defined as those non-manmade objects in space whose orbits bring them within a set distance of the Earth generally equivalent to approximately 50 million kilometers (31 million miles), with a portion of these objects traveling sufficiently closer to make an eventual collision a possibility. The larger NEOs (those with a maximum physical dimension of more than a meter[1]) are generally referred to as either asteroids or comets, while smaller objects are referred to as meteoroids. Large or small, they are all called meteors upon fiery transit of Earth's atmosphere. When pieces of a meteor survive transit through the atmosphere to strike the surface of the Earth, they are called meteorites.

Every day, a continual influx of these objects strikes Earth's atmosphere. Most of them are dust-sized particles, but they add up; it's been estimated that on a typical day, these particles total from 50 to 150 tons of matter. Asteroids of the order of a few meters in size strike the atmosphere roughly annually. Damage on Earth's surface is likely only when the kinetic energy of the object – the energy it carries by virtue of its mass and velocity when it enters the atmosphere – is in the range of a few hundred kilotons of TNT equivalent or above. (By comparison, the Hiroshima atomic bomb was roughly 13 kilotons of TNT equivalent.) Asteroids with this much energy are thought to strike the Earth only every 100 years or so. The more frequent, less energetic ones generally deposit that energy high enough in the atmosphere that no effects are felt at Earth's surface.

Asteroids can be divided into three broad categories on the basis of their composition: carbon-aceous, stony, and metallic. Metallic asteroids are denser than the other varieties and therefore have greater destructive power, being both more massive for a given size and less likely to disintegrate during atmospheric entry. Stony asteroids are the most common variety; at a typical

---

[1] Only the largest asteroids — those few with diameters in the range of hundreds of kilometers — are spherical. Smaller ones are of irregular shape.

approach velocity of 16 kilometers per second[2] (36,000 miles per hour), a stony asteroid would have to be around 20 meters in size to deliver the few hundred kilotons of TNT equivalent needed to do significant damage at the surface. (The asteroid that exploded over Chelyabinsk on February 15 is estimated to have been about 17 meters in size.)

The 1908 explosion over Tunguska in Siberia, which leveled trees over an area of more than 2,200 square kilometers (850 square miles), is thought to have been caused by a stony asteroid between 45 and 60 meters in diameter, imparting between 10 and 20 megatons of TNT equivalent. Such a strike is believed to be a once-in-a-thousand-year event (or, put differently, having a 1 in 1,000 chance of occurring in any given year). An energy release of this size could cause hundreds of thousands of casualties and massive destruction if it occurred over an urban area. The probability of this occurring, however, is much smaller than the one-in-one-thousand odds I just mentioned, inasmuch as land covers only 30 percent of the area of the planet and cities only 2-3 percent of the land area. Of course, a similarly sized or even larger asteroid that made it to the surface intact could cause significant damage even if it hit the ocean, by virtue of the resulting tsunami.

Depending on its composition and velocity, an asteroid of 140 meters in diameter could have an impact energy in the range of 50 to 500 megatons of TNT equivalent and would be capable of causing destruction over a large region. The probability of a strike of this magnitude has been estimated at about 1 in 30,000 per year. As I noted earlier, it is believed that more than 90 percent of all NEOs 1 kilometer or greater in diameter have now been identified. None of those identified so far appears to pose a risk of collision, which is fortunate as the impact of an object of this size would release between 20,000 and 200,000 megatons of TNT equivalent and likely imperil all of civilization. The object that impacted the Earth just off the Yucatan Peninsula 65 million years ago, and that is believed to have been responsible for mass extinctions across the planet, was an asteroid estimated to have been 10 kilometers in diameter.

RECENT EXAMPLES

In recent years, several NEOs have made close passes by Earth. Two unrelated asteroids, estimated to have been between six and 20 meters in diameter, passed between the Earth and the Moon on September 8th, 2010. Neither posed a risk of striking the Earth, but they served as a reminder that these kinds of close flybys are not rare. It's estimated that almost every day, at least one 10-meter near-Earth asteroid (part of the undiscovered population of about 50 million) passes the Earth inside the orbit of the Moon.

The Chelyabinsk asteroid strike of February 15 lit up the sky and unleashed a series of shock waves that shattered windows over a wide area. Approximately 1,500 people were injured and 60 were hospitalized; the regional governor said that two-thirds of the injuries were light wounds from shattered glass and other materials. More than 4,000 buildings were damaged. According to NASA estimates, the velocity of approach of the incoming asteroid was about 18 kilometers per second and its energy was about 400 kilotons of TNT equivalent. These estimates enable us to calculate a corresponding mass of 11,000 metric tons. Early measurements in the hours

---

[2] Most asteroids with orbits crossing that of the Earth would impact with velocities between 12 and 20 kilometers per second, hence the choice of 16 kilometers per second as an intermediate figure.

following the event underestimated the energy (and therefore the mass) of the asteroid, but with the addition of data in the following days both from the ground and from orbiting satellites that witnessed the asteroid's entry, the energy measurement was refined. The entire event from the asteroid's atmospheric entry to its disintegration at an altitude of about 20 kilometers took only 32.5 seconds. Thus far, approximately 50 small meteorites resulting from this explosion have been found. One piece blasted a six-meter-diameter hole in the ice covering one of Siberia's many lakes.

On the same day as the unexpected Chelyabinsk event, the 45-meter asteroid 2012 DA14 safely passed nearly 27,700 kilometers (17,200 miles) from Earth, a close flyby that had been predicted many months in advance. This asteroid had been tracked since its February 2012 discovery. Tracking data and orbit calculations made by the various Federal agencies with responsibilities in this area made it clear over a year ago that this asteroid did not pose a threat to Earth, the International Space Station, or satellites in orbit. This event allowed researchers to measure the object's path and orbit with greater precision, improving estimates of future near-Earth passes.

Analysis of data collected during these two contemporaneous events indicates that the asteroid that exploded over Chelyabinsk was almost certainly unrelated to the larger asteroid 2012 DA14; the smaller asteroid's trajectory was not consistent with its being a fragment that came off the larger one. It's notable that while the Chelyabinsk "fireball" was unusual for its size and visibility, thousands of smaller strikes that still explode with enough energy to produce a fireball (defined as being brighter than the brightest planet) occur each day. Most fireballs are not noticed because they occur over oceans or uninhabited regions or are masked by daylight. Nearly all of them are caused by objects too small to be detected before they enter Earth's atmosphere… and also too small to do damage on Earth's surface.

DETECTION AND MITIGATION EFFORTS

The ability to detect NEOs and to determine whether a collision with Earth is likely depends on the distance, size, and reflectivity of the object and the number and capabilities of the telescopes that are looking for it. In general, detection of NEOs and prediction of future orbits are challenging endeavors, especially when one considers that orbits can change as a result of encounters with other objects.

NASA sponsors a number of activities relating to the search for NEOs under its Near Earth Object Observation (NEOO) program, including work at the international Minor Planet Center (MPC), located at the Harvard-Smithsonian Center for Astrophysics, which collects and correlates NEO orbit data; research at two radio-telescope facilities that help provide precision tracking and characterization of NEOs; surveys conducted by ground-based optical telescopes; and activities at the NASA NEO Program Office at the Jet Propulsion Laboratory (JPL), which coordinates assessments of NEO orbits and impact probabilities. There are also cooperative projects involving NASA, the National Science Foundation (NSF, which has a key role within the United States for ground-based astronomical assets), and the U.S. Air Force (USAF) Panoramic Survey Telescope and Rapid Response System (PanSTARRS) program, as well as non-government academic and space research organizations. Additionally, NEO detection is a major science driver for the proposed Large Synoptic Survey Telescope. NASA is also working with the Canadian Space Agency (CSA) on processing of data that will be collected from the CSA Near-Earth Object Surveillance Satellite (NEOSSat) launched last week (February 25).

The Administration places a high priority on tracking asteroids and protecting our planet from them, as evidenced by the five-fold increase in the budget for NASA's NEOO program since 2009. The United States has an effective program for discovering larger NEOs, but we need to improve our capabilities for the identification and characterization of smaller NEOs. Specifically, with our current or near-future capabilities, both on the ground and in space, it is unlikely that objects smaller than 100 meters in diameter on collision courses with the Earth will be detected with greater than weeks of advance warning – a matter of some concern since the larger objects in this range could be city-destroyers.

ADMINISTRATION POLICIES AND BUDGETS

Finally, I'd like to underscore some Administration policy and budgetary decisions relating to NEOs, which will buttress ongoing detection and tracking activities going forward.

The President's 2013 Budget for NASA's NEOO Program proposes more than a five-fold increase in funding (to $20.5 million from $4 million) from the 2009 funding level for NEO detection activities. Further, the President's National Space Policy specifically directs NASA to "pursue capabilities, in cooperation with other departments, agencies, and commercial partners, to detect, track, catalog, and characterize near Earth objects to reduce the risk of harm to humans from an unexpected impact on our planet and to identify potentially resource-rich planetary objects." This guidance also reinforces NASA's roles and responsibilities with regard to NEOs, as well as those of other Federal departments and agencies including the Department of Defense, the Department of State, and the Department of Homeland Security's Federal Emergency Management Agency (FEMA).

In furtherance of these directives, NASA has completed a number of missions to investigate asteroids and has others planned. For example, the Near Earth Asteroid Rendezvous (NEAR)-Shoemaker mission rendezvoused, orbited, and touched down on the near-Earth asteroid Eros in 2001, significantly advancing the field of asteroid studies. The United States also collaborated with Japan, through NASA, on the successful asteroid visit and sample return mission known as Hayabusa. The OSIRIS-REx mission, currently in development for launch in 2016, will study, characterize, and return to Earth a sample of near-Earth asteroid 1999 RQ36 in an effort to investigate planet formation and the origin of life. And of course NASA is committed to carrying out the President's goal of conducting a human mission to an asteroid by 2025. That mission will benefit from current efforts to detect, track, and characterize NEOs by speeding the identification of potential targets for exploration. And in return, such a mission will generate invaluable information for use in future detection and mitigation efforts.

OSTP has been working closely with several departments and agencies to draft plans and procedures, including potential mitigation strategies, that could be used in the unlikely event of a NEO impact threat. Under these plans, it is NASA's responsibility to provide initial notice of such a threat. Following such notification, communications resources and mechanisms already in place within FEMA would be used to communicate information domestically. The Department of State's diplomatic mechanisms would come into play for international communications as needed.

With regard to risk mitigation, the Administration is committed to exploring and developing the capabilities and techniques necessary to protect the Earth in general, and the United States in particular, from NEO threats and implementing a collision mitigation campaign if necessary and appropriate. In 2008, the Executive Office of the President collaborated with NASA and the U.S. Air Force to run the first-ever disaster and deflection exercise, which included members from the National Security Staff, Joint Chiefs of Staff, Office of the Under Secretary of Defense for Policy, Missile Defense Agency, the Defense Threat Reduction Agency, the Department of Defense (DoD) National Security Space Office (now the Executive Agent for Space Staff), and the Department of Energy. The National Research Council and the NASA Advisory Council have also provided helpful recommendations and guidance on research priorities in impact mitigation techniques.

Among the highlighted needs are an improved understanding of NEO characteristics to enable more refined impact experiments; enhanced computer simulations; crewed and uncrewed *in situ* asteroid investigations; and further research and capabilities development in the domain of deflection, including explosive technologies, and impact scenarios (including design reference missions and gaming exercises). DoD and NASA have already shown tremendous leadership by taking the initiative to run multi-agency disaster and deflection exercises, and by collaborating in the development of an international disaster and deflection response scenario for the upcoming Planetary Defense Conference hosted by the International Academy of Astronautics in Flagstaff, Arizona.

In summary, the Administration, with the support of Congress, has taken many positive steps to improve NEO detection capabilities, including meeting the 90 percent detection goal for one-kilometer asteroids. Much more needs to be done, however, and it is important to note that a challenge on the scale of planetary defense cannot be met by any single nation or government alone. Rather it will be critical going forward that the Federal Government cooperate closely with domestic partners in industry, academia, and other sectors as well as with foreign governments and international organizations to achieve our shared goal of scientific discovery, exploration, and risk mitigation.

I thank the Committee for its continued support and interest in this issue and I look forward to continuing to work with you on it. I will be pleased to take any questions Members may have.

DR. JOHN P. HOLDREN is Assistant to the President for Science and Technology and Director of the White House Office of Science and Technology Policy. Trained in aerospace engineering and theoretical plasma physics at MIT and Stanford, he is a member of the National Academy of Sciences, the National Academy of Engineering, and the American Academy of Arts and Sciences, as well as a foreign member of the Royal Society of London and a former President of the American Association for the Advancement of Science. Prior to joining the Obama administration, he was a professor in both the Kennedy School of Government and the Department of Earth and Planetary Sciences at Harvard, as well as Director of the Woods Hole Research Center. From 1973 to 1996 he was on the faculty of the University of California, Berkeley, where he co-founded and co-led the interdisciplinary graduate-degree program in energy and resources.

Chairman SMITH. Thank you, Dr. Holdren.
General Shelton.

**TESTIMONY OF GEN. WILLIAM L. SHELTON,
COMMANDER, U.S. AIR FORCE SPACE COMMAND**

General SHELTON. Mr. Chairman, Representative Johnson and distinguished Members of the Committee, it is an honor to appear before you today. It is also a privilege to appear with my colleagues and teammates in the space community.

Space situational awareness underpins our entire spectrum of space activities, and Air Force Space Command is proud of our crucial role in monitoring activity in the space domain. Specifically, we provide capabilities employed ultimately by United States Strategic Command to detect, track, identify and characterize human-made objects in Earth orbit. Our sensors also are capable of detecting natural phenomena like bolides.

However, the Nation's current capability to track asteroids is dependent upon NASA and other organizations such as the Massachusetts Institute of Technology's Lincoln Laboratory. For example, during the recent asteroid 2012 DA14 event, the Joint Space Operations Center at Vandenberg Air Force Base in California used tracking data from NASA's Near Earth Object Program Office at the Jet Propulsion Laboratory to perform collision avoidance screenings to ensure the safety of our satellites. We remain committed to working closely with our partners to ensure comprehensive space situational awareness for the Nation.

I thank you for the opportunity to appear before you, and I look forward to your questions.

[The prepared statement of General Shelton follows:]

NOT FOR PUBLICATION UNTIL RELEASED BY THE
SCIENCE, SPACE AND TECHNOLOGY COMMITTEE
U.S. HOUSE OF REPRESENTATIVES

DEPARTMENT OF THE AIR FORCE

PRESENTATION TO THE

HOUSE SCIENCE, SPACE AND TECHNOLOGY COMMITTEE

U.S. HOUSE OF REPRESENTATIVES

SUBJECT: THREATS FROM SPACE: A REVIEW OF U.S. GOVERNMENT EFFORTS TO TRACK AND MITIGATE ASTEROIDS AND METEORS

STATEMENT OF: GENERAL WILLIAM L. SHELTON
COMMANDER, AIR FORCE SPACE COMMAND

March 19, 2013

NOT FOR PUBLICATION UNTIL RELEASED BY THE
SCIENCE, SPACE AND TECHNOLOGY COMMITTEE
U.S. HOUSE OF REPRESENTATIVES

Chairman Smith, Ranking Member Johnson, and Members of the Committee, thank you for the opportunity to appear before you today to discuss Air Force Space Command's role in monitoring activity in the space domain. Space situational awareness underpins the entire spectrum of space activities and Air Force Space Command's focus is on providing forces and capabilities to United States Strategic Command (USSTRATCOM) to detect, track, identify and characterize human-made objects that orbit the Earth. Our efforts contribute to the collaborative, multiagency endeavor required to ensure comprehensive space situational awareness for the Nation.

Air Force Space Command Roles and Responsibilities

Air Force Space Command presents space forces and capabilities to USSTRATCOM through the Fourteenth Air Force. The Commander, Fourteenth Air Force, Lieutenant General Susan Helms, is dual-hatted as the Commander, Joint Functional Component Command for Space (JFCC SPACE) and is responsible for executing USSTRATCOM's space operations mission.

JFCC SPACE's Joint Space Operations Center (JSpOC) is the avenue through which JFCC SPACE commands and controls space forces and it is the epicenter of the space situational awareness mission. The JSpOC is also the means by which JFCC SPACE coordinates space situational awareness with other agencies. For example, National Aeronautics and Space Administration (NASA) orbital safety analysts reside within the JSpOC 24 hours a day to collaborate on orbital safety threats to human space flight.

Detecting and Tracking Human-Made Objects in Space

For definitional purposes, the Air Force considers an object as being near-Earth if it takes less than 225 minutes to complete its orbit around the Earth. That is roughly 5,800 kilometers in altitude. All else is characterized as deep-space.

All entities that operate in the space domain are increasingly concerned about orbital debris. Past practices, as well as recent events, both accidental and purposeful, have created a troublesome debris environment in low Earth orbit. In 2007, the People's Republic of China performed an anti-satellite test which successfully struck its target, one of their defunct weather satellites. In 2009, an active Iridium communications satellite and a non-operative Russian Cosmos satellite accidentally collided. Each of these near-Earth incidents resulted in thousands of pieces of debris large enough to track. As of March 1, 2013, we continue to track 2,160 pieces from the Iridium-Cosmos collision alone. Additionally, in 2012, a Russian BRIZ-M upper stage malfunctioned with a significant quantity of propellant remaining. This upper stage eventually exploded and we are now tracking almost 150 pieces of debris from this event. More troubling is that our modeling tells us that each event produced thousands more pieces of debris which are too small for our sensors to reliably track. At orbital velocities, these small objects still represent catastrophic potential threats to fragile spacecraft. Each subsequent collision, explosion or break-up leaves more debris in space, increasing the potential for further collisions and even more debris, a chain reaction that could exponentially increase the risk to activities in space.

To support national security space operations in an environment of increasingly adverse environmental conditions, the JSpOC collects and processes data from a worldwide network of radar and optical sensors, as well as a dedicated space surveillance satellite. Each day the JSpOC creates and disseminates over 200,000 sensor taskings. The sensors then return nearly 500,000 observations to the JSpOC for processing. JSpOC operators use this data to maintain a high accuracy catalog of space objects and perform over 1,000 satellite collision avoidance screenings daily. These operations form the basis of the United States' space situational awareness capability, which is then shared with other operators in the national security, civil and commercial sectors of space operations.

Size of Objects and Distance Detected

The JSpOC directs space surveillance sensors to track objects in space ranging in size from as small as a softball to as large as the International Space Station. Today, the JSpOC tracks approximately 23,000 objects in both near-Earth and deep-space, but there is an estimated half million plus human-made objects in Earth orbit that we are not tracking. The number of objects reliably catalogued by the JSpOC is expected to rise by as much as four or five times by 2030 due to the steady growth in the on-orbit population, as well as the planned fielding of improved sensor capability.

The systems Air Force Space Command supports were designed and fielded to meet requirements specific to national security missions. Some track objects in near-Earth orbit while others are focused on deep-space, primarily geosynchronous orbit; however, we can on occasion support other orbit profiles. For example, when the NASA Stardust spacecraft returned to Earth from collecting samples of the coma of a comet in 2006, we were able to modify certain parameters of existing models and use space surveillance sensors to track the Stardust sample return capsule in its parabolic return-to-Earth.

Technologies and Processes

As previously stated, Air Force Space Command sensors were developed to track man-made objects in Earth orbit. The Nation's current capability to track asteroids, which orbit the sun, is largely driven by NASA. Air Force developmental telescopes are used by the Massachusetts Institute of Technology's Lincoln Laboratory to find and catalog asteroids under contract to NASA. And in some cases, the JSpOC can task space surveillance sensors to help track close approaches by asteroids and help predict potential collisions with Earth-orbiting objects. For example, during the recent Asteroid 2012 DA14 event, the JSpOC used orbit data for it from NASA's Near Earth Object Program Office at the Jet Propulsion Laboratory to screen for potential collisions with man-made objects in Earth orbit.

The current sensor tasking and data processing system used by the JSpOC to accomplish the space situational awareness mission was designed in the 1980s, fielded in the early 1990s and is nearing its capacity limits and end-of-life. We are in the process of fielding the next generation system, the JSpOC Mission System (JMS). With its open, service-oriented architecture, the JMS will supply the automation necessary to make better use of the tremendous volume of sensor data available. It will also enhance the Commander, JFCC SPACE capability to conduct space operations in a much more efficient and much safer manner.

Conclusion

Space situational awareness is foundational to civil, military and commercial space activities. Air Force Space Command forces and capabilities to detect, track, identify and characterize man-made objects in Earth orbit support the larger collaborative effort to maintain space situational awareness for the Nation.

GENERAL WILLIAM L. SHELTON

BIOGRAPHY
UNITED STATES AIR FORCE

GENERAL WILLIAM L. SHELTON

Gen. William L. Shelton is Commander, Air Force Space Command, Peterson Air Force Base, Colo. He is responsible for organizing, equipping, training and maintaining mission-ready space and cyberspace forces and capabilities for North American Aerospace Defense Command, U.S. Strategic Command and other combatant commands around the world. General Shelton oversees Air Force network operations; manages a global network of satellite command and control, communications, missile warning and space launch facilities; and is responsible for space system development and acquisition. He leads more than 42,000 professionals assigned to 134 locations worldwide.

General Shelton entered the Air Force in 1976 as a graduate of the U.S. Air Force Academy. He has served in various assignments, including research and development testing, space operations and staff work. The general has commanded at the squadron, group, wing and numbered air force levels, and served on the staffs at major command headquarters, Air Force headquarters and the Office of the Secretary of Defense. Prior to assuming his current position, General Shelton was the Assistant Vice Chief of Staff and Director, Air Staff, U.S. Air Force, the Pentagon, Washington, D.C.

EDUCATION
1976 Bachelor of Science degree in astronautical engineering, U.S. Air Force Academy, Colorado Springs, Colo.
1980 Master of Science degree in astronautical engineering, U.S. Air Force Institute of Technology, Wright-Patterson AFB, Ohio
1986 Armed Forces Staff College, Norfolk, Va.
1995 Master of Science degree in national security strategy, National War College, Fort Lesley J. McNair, Washington, D.C.
1996 Program for Senior Officials in National Security, Syracuse University and Johns Hopkins University
1997 Fellow, Seminar XXI, Massachusetts Institute of Technology, Cambridge

ASSIGNMENTS
1. August 1976 - May 1979, Launch Facilities Manager, Launch Director and Technical Assistant to the Commander, Space and Missile Test Center, Vandenberg AFB, Calif.
2. May 1979 - December 1980, Graduate Student, U.S. Air Force Institute of Technology, Wright-Patterson AFB, Ohio
3. January 1981 - July 1985, Space Shuttle Flight Controller, Johnson Space Center, Houston, Texas
4. July 1985 - January 1986, Student, Armed Forces Staff College, Norfolk, Va.
5. January 1986 - July 1988, Staff officer, Deputy Chief of Staff for Operations, Air Force Space Command, Peterson AFB, Colo.
6. August 1988 - August 1990, Staff Officer, Office of Space Plans and Policy, Office of the Secretary of the Air Force, Washington, D.C.
7. August 1990 - June 1992, Commander, 2nd Space Operations Squadron, Falcon AFB, Colo.
8. June 1992 - June 1993, Executive Officer to the Vice Commander, Air Force Space Command, Peterson AFB, Colo.
9. June 1993 - July 1994, Commander, 50th Operations Group, Falcon AFB, Colo.
10. August 1994 - June 1995, Student, National War College, Fort Lesley J. McNair, Washington, D.C.

GENERAL WILLIAM L. SHELTON

11. June 1995 - September 1997, Deputy Program Manager and Executive Assistant, Cooperative Threat Reduction Program Office, Office of the Assistant to the Secretary of Defense for Nuclear, Chemical and Biological Defense Programs, Washington, D.C.
12. September 1997 - August 1999, Commander, 90th Space Wing, Francis E. Warren AFB, Wyo.
13. September 1999 - July 2000, Chief, Space Superiority Division, Office of the Deputy Chief of Staff for Plans and Programs, Headquarters U.S. Air Force, Washington, D.C.
14. July 2000 - November 2000, Director of Manpower and Organization, Office of the Deputy Chief of Staff for Plans and Programs, Headquarters U.S. Air Force, Washington, D.C.
15. November 2000 - May 2002, Director of Requirements, Headquarters Air Force Space Command, Peterson AFB, Colo.
16. June 2002 - January 2003, Director of Plans and Programs, Headquarters AFSPC, Peterson AFB, Colo.
17. January 2003 - May 2003, Director, Air and Space Operations, Headquarters AFSPC, Peterson AFB, Colo.
18. June 2003 - January 2005, Director of Capability and Resource Integration (J8), USSTRATCOM, Offutt AFB, Neb.
19. January 2005 - May 2005, Director of Plans and Policy (J5), USSTRATCOM, Offutt AFB, Neb.
20. May 2005 - December 2008, Commander, 14th Air Force (Air Forces Strategic), AFSPC, and Commander, Joint Functional Component Command for Space, USSTRATCOM, Vandenberg AFB, Calif.
21. December 2008 - July 2009, Chief of Warfighting Integration and Chief Information Officer, Office of the Secretary of the Air Force, the Pentagon, Washington, D.C.
22. July 2009 - January 2011, Assistant Vice Chief of Staff and Director, Air Staff, U.S. Air Force, the Pentagon, Washington, D.C.
23. January 2011 - present, Commander, Air Force Space Command, Peterson AFB, Colo.

SUMMARY OF JOINT ASSIGNMENTS
1. June 1995 - September 1997, Deputy Program Manager and Executive Assistant, Cooperative Threat Reduction Program Office, Office of the Assistant to the Secretary of Defense for Nuclear, Chemical and Biological Defense Programs, Washington, D.C., as a colonel
2. June 2003 - January 2005, Director of Capability and Resource Integration (J8), USSTRATCOM, Offutt AFB, Neb., as a brigadier general and major general
3. January 2005 - May 2005, Director of Plans and Policy (J5), USSTRATCOM, Offutt AFB, Neb., as a major general
4. May 2005 - July 2006, Commander, Joint Space Operations, USSTRATCOM, Vandenberg AFB, Calif., as a major general
5. July 2006 - December 2008, Commander, Joint Functional Component Command for Space, USSTRATCOM, Vandenberg AFB, Calif., as a major general and lieutenant general

BADGES
Command Space Badge
Master Cyberspace Badge
Parachutist Badge

MAJOR AWARDS AND DECORATIONS
Distinguished Service Medal with oak leaf cluster
Defense Superior Service Medal with oak leaf cluster
Legion of Merit with oak leaf cluster
Defense Meritorious Service Medal with oak leaf cluster
Meritorious Service Medal with four oak leaf clusters
Air Force Commendation Medal
Joint Meritorious Unit Award with two oak leaf clusters
Air Force Outstanding Unit Award with silver and two bronze oak leaf clusters
Air Force Organizational Excellence Award with oak leaf cluster

EFFECTIVE DATES OF PROMOTION
Second Lieutenant June 2, 1976
First Lieutenant June 2, 1978
Captain June 2, 1980
Major May 1, 1985
Lieutenant Colonel March 1, 1990
Colonel Feb. 1, 1994
Brigadier General Jan. 1, 2001
Major General July 1, 2004
Lieutenant General Dec. 20, 2007
General Jan. 5, 2011

(Current as of January 2013)

Chairman SMITH. Thank you, General Shelton.
Administrator Bolden.

TESTIMONY OF THE HON. CHARLES F. BOLDEN, JR., ADMINISTRATOR, NATIONAL AERONAUTICS AND SPACE ADMINISTRATION

General BOLDEN. Mr. Chairman and Members of the Committee, thank you for the opportunity also to appear today to discuss the topic of near-Earth objects, and before I formally begin, Mr. Chairman, I would like to congratulate you on your appointment as the new Chairman of the House Science, Space, and Technology Committee, and I look forward to working with you in that capacity.

I would also like to thank you, Mr. Chairman, and Congresswoman Edwards and Congressman Holt, who is not here, for the recent op-eds that you wrote that called more attention to this for the American public, which I think is really important.

The events of February 15, 2013, were a stark reminder of why NASA has for years devoted a great deal of attention to near-Earth objects and why this hearing is so timely and important. The events of February 15 also highlight the wisdom of Congress, the Administration and NASA in enabling a human exploration of an asteroid.

The predicted close approach of a small asteroid called 2012 DA14 and the unpredicted entry and explosion of a very small asteroid about 15 miles above Russia that Dr. Holdren talked about earlier have focused a great deal of public attention on the necessity of tracking asteroids and other near-Earth objects and protecting our planet from them, something this Committee and NASA have been working on for over 15 years. Again, NASA has been focused on tracking asteroids and protecting our home planet from them well before these recent events. In fact, NASA's focus in this area is evident from our fivefold increase in near-Earth object budgets since 2010, and literally dozens of people are involved with some aspect of our NEO research across NASA and its field centers.

In addition to the resources NASA puts into understanding asteroids, the agency partners with university astronomers, space science institutes and other agencies across the country that are working to track and better understand these near-Earth objects, often with grants, interagency transfers and other contracts from NASA.

The new public attention is not hard to understand. The coincidence of having these two very rare events happening on the same day along with the unfortunate injuries of over 1,000 people on the ground in Russia made this a very big news event. However, we should remember that the probability of any sizable NEO impacting the Earth any time in the next 100 years is extremely remote.

To put these two recent events in context, very small objects enter the Earth's atmosphere all the time. Current estimates are that on average, about 80 tons of material in the form of dust grains and small meteoroids enter the Earth's atmosphere every single day, objects the size of a basketball arrive once a day, and objects as large as a car arrive about once per week. Our Earth's atmosphere protects us from these small objects, so nearly all are destroyed before hitting the ground and pose no threat to life here on Earth. However, the potential consequences of a significant impact are potentially very great indeed. Consistent with NASA's role as established by Congress and prescribed in the President's National Space Policy, NASA has taken a leadership role to pursue capabilities to detect, track and characterize near-Earth objects to reduce the risk of harm to humans from an unexpected impact on our planet.

NASA is also developing new vehicles and capabilities including Orion and the Multipurpose Crew Vehicle and the Space Launch System, which will enable human exploration of the solar system beyond low-Earth orbit. As the President stated in his April 15, 2010, speech at the Kennedy Space Center, NASA's intention is to "send astronauts to an asteroid for the first time in history" and NASA is working to accomplish this mission by 2025. In fact, NASA leads the world in the detection and characterization of NEOs and is responsible for the discovery of about 98 percent of all known NEOs.

Now, here I will take a risk. There should be a chart coming up very soon. It is. Thank you. As shown in this graphic, the cumulative discovery of near-Earth asteroids started picking up dramatically in 1998 with the start of NASA's Spaceguard Search program, and the number of known near-Earth asteroids has grown from a few hundred to nearly 10,000 in just 15 years, and I think it is not insignificant that it goes almost asymptotic when you look at 2005 when the Congress, NASA and the Administration really picked up the emphasis on that.

NASA continues to make progress toward the goals set for us by the Congress. To date, over 9,600 near-Earth asteroids of all sizes have been found. Larger asteroids pose a greater threat to the planet as a whole, and the percentage of asteroids we have identified tracks this relationship. We found 95 percent of the largest NEOs over 1 kilometer in size. Our current estimate is that we have also found about 60 percent of the NEOs that are between 300 meters and 1 kilometer. As the graphic shows, we still have some work to do to find NEOs in the 140-meter class, and the next graphic please. You can see here the total discovered per size and you can see where we are lacking as the sizes go down.

Our remote ground-based observations of comets and asteroids have been augmented by close-up reconnaissance data from our science missions. From 1997 to 2001, NASA's near-Earth asteroid rendezvous flyby flew by two main asteroid belts before orbiting and landing on the near-Earth asteroid 433 Eros. Last August, our Dawn spacecraft departed the asteroid Vesta and is now on its way to a 2015 rendezvous with Ceres, the solar system's largest asteroid. Launching in 2016, NASA's OSIRIS–REx mission will return a sample of up to 2.2 pounds from an asteroid to Earth in 2023.

Of course, NASA is working to accomplish an astronaut visit to an asteroid by 2025. This mission and the vital precursor activities that will be necessary to ensure its success should result in additional insight into the nature and composition of NEOs and will increase our capability to approach and interact with asteroids.

NASA has a long history of observing comets and asteroids but as their importance as potentially hazardous objects has become apparent, NASA has significantly increased its program of detection, reconnaissance and characterization. We have gained a nearly complete understanding of the population of NEOs over 1 kilometer in size, and we are making marked progress in protecting our planet from smaller but still dangerous objects. While we emphasize that the risks form impacts are remote, we remain absolutely committed to fulfilling our responsibility to find and track near-Earth objects. We will continue to scan the skies and update the Congress and the world on what we find.

Again, thank you very much for the opportunity to testify today, and I look forward to responding to any questions you may have.

[The prepared statement of General Bolden follows:]

HOLD FOR RELEASE
UNTIL PRESENTED
BY WITNESS
March 19, 2013

**Statement of
Mr. Charles Bolden
Administrator
National Aeronautics and Space Administration**

before the

**Committee on Science, Space, and Technology
U.S. House of Representatives**

Mr. Chairman and Members of the Committee, thank you for the opportunity to appear today to discuss the topic of Near Earth-Objects (NEOs). NEOs are defined as those non-manmade objects in space whose orbits bring them within 1.3 Astronomical Units of the Sun, or to a set distance to the Earth's orbit that is generally equivalent to approximately 50 million kilometers (31 million miles).

The events of February 15, 2013 were a stark reminder of why NASA has for years devoted a great deal of attention to Near Earth Objects and why this hearing is so timely and important. The predicted close approach of a small asteroid, called 2012 DA14, and the unpredicted entry and explosion of a very small asteroid about 15 miles above Russia, have focused a great deal of public attention on the necessity of tracking asteroids and other NEOs and protecting our planet from them -- something this Committee and NASA have been working on for over 15 years. The events of February 15th also highlight the wisdom of the Congress, the Administration and NASA in enabling the human exploration of an asteroid.

The new public attention is not hard to understand. The coincidence of having these two very rare events, happening on the same day, along with the unfortunate injuries to over 1,000 people on the ground in Russia, made this a very big news story. However, we should remember that the probability of any sizable NEO impacting the Earth anytime in the next 100 years is extremely remote. The small fraction of objects we have discovered which do have the potential to impact the Earth are tagged by the Minor Planet Center as Potentially Hazardous Objects (or PHOs) and are subjected to further study and observation to assess just how hazardous they may be. Smaller objects, such as the recent impact in Russia, will always be difficult to detect and provide adequate warning; however, progress will be made over the next decade in this area as well.

To put these two recent events in context, very small objects enter the Earth's atmosphere all the time. The larger NEOs (those with a maximum physical dimension of more than a meter) are generally referred to as either asteroids or comets, while smaller objects are referred to as meteoroids. Current estimates are that on average about 100 tons of material in the form of dust grains and small meteoroids enter the Earth's atmosphere each day. Objects the size of a basketball arrive about once per day, and objects as large as a car arrive about once per week. Our Earth's atmosphere protects us from these small objects, so nearly all are destroyed before hitting the ground and generally pose no threat to life on Earth.

However, while objects the size of the one that exploded over Russia, which we have assessed as a rocky asteroid about 17 meters in diameter and weighing from 7,000 to 13,000 metric tons, enter the Earth's atmosphere very rarely on human timescales, they do have serious consequences. NASA has been at the forefront in leveraging our own resources, as well as interagency, international, academic and commercial partnerships, to study both these rare and more common NEO close approach events, and to expand our knowledge about NEOs and the potential threat they pose to the Earth. The 2010 National Space Policy specifically directs NASA to take a leadership role to "pursue capabilities, in cooperation with other departments, agencies, and commercial partners, to detect, track, catalog, and characterize near-Earth objects to reduce the risk of harm to humans from an unexpected impact on our planet and to identify potentially resource-rich planetary objects."

NASA leads the world in the detection and characterization of NEOs, and is responsible for the discovery of about 98 percent of all known NEOs. Over 15-plus years of collecting data on NEOs has helped to shape the scientific consensus about these objects and the potential threat they pose to the Earth. NASA is leading a wide array of activities related to NEOs, including a long-standing ground-based observing campaign, focused flight missions to study both asteroids and comets, as well as conceptual studies and technology development to improve our ability to find NEOs. NASA uses radar techniques to better characterize the orbits, shapes, and sizes of observable NEOs, and funds research activities to better understand their composition and nature. NASA also funds the key reporting and dissemination infrastructure that allows for world-wide follow-up observations of NEOs as well as research related activities, including computer modeling, sample analysis and workshops to disseminate information about NEOs to the larger scientific and engineering community. Consistent with NASA's role as outlined in President's National Space Policy, NASA continues to collaborate with the Executive Office of the President and the Department of Defense on planning and exercises for responding to future hazardous NEOs.

NASA is also developing new vehicles and capabilities, including the Orion Multi-Purpose Crew Vehicle and the Space Launch System, which will enable human exploration of the solar system beyond low Earth orbit. As the President stated in his April 15, 2010, speech at the Kennedy Space Center, NASA's intention is to "[send] astronauts to an asteroid for the first time in history." NASA is working to accomplish this mission by 2025. This mission, and the vital precursor activities that will be necessary to ensure its success, should result in additional insight into the nature and composition of NEOs and will increase our capability to approach and interact with asteroids.

Detection-related Activities
NASA was tasked by Congress in 1998 to catalog 90 percent of all the large NEOs (those of 1 kilometer or more in size) within 10 years; these would be large enough that should they strike Earth, it would result in a global catastrophe. NASA worked with a number of ground-based observatories and partners as part of our Spaceguard survey to reach that goal; NASA has now catalogued an estimated 95 percent of all NEOs over 1 km in size. None of these known large NEOs pose any threat of impact to the Earth anytime in the foreseeable future.

As shown in the next graphic, the cumulative discovery of Near-Earth Asteroids, the largest subset of NEOs, started picking up dramatically in 1998 with the start of NASA's Spaceguard search program and the number of known NEOs has grown from a few hundred to nearly 10,000 in just 15 years.

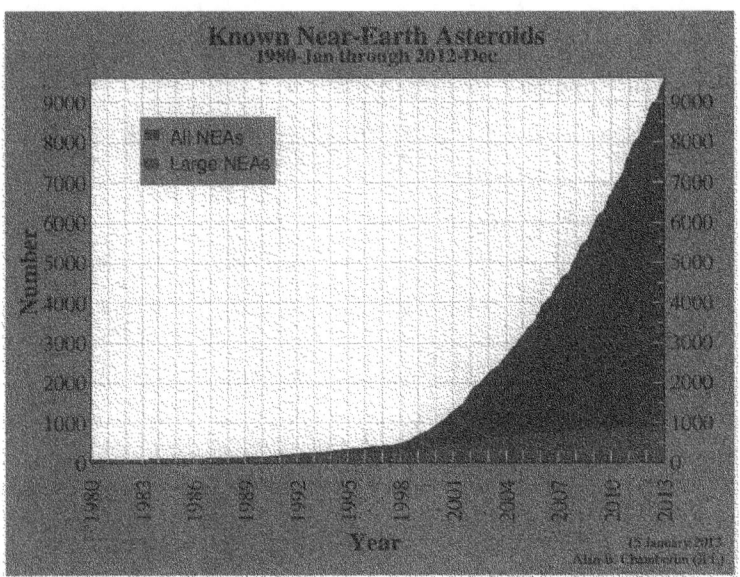

In 2005, as part of the NASA Authorization Act of 2005 (P.L. 109-155), Congress also directed NASA to initiate the George E. Brown, Jr. Near-Earth Object Survey, to "detect, track, catalogue, and characterize the physical characteristics of near-Earth objects equal to or greater than 140 meters in diameter" and set a goal for this program to achieve 90 percent completion by 2020. That effort is underway. NASA's NEO Observation Program currently funds three survey teams that operate five ground-based telescopes involved in the NEO search effort. Each team conducts independent operations for 14 to 20 nights per month, as weather permits, avoiding approximately a week on either side of the full moon when the sky is too bright to detect these extremely dim objects from the ground.

NASA also leveraged its investment in the Wide-field Infrared Survey Explorer (WISE) spacecraft to discover NEOs. WISE was designed originally as an astrophysics mission that would scan the entire sky in infrared light to study the coolest stars and the universe's most luminous galaxies. After completion of its astrophysics mission, WISE's operations were extended specifically as a NEO-finding mission. NASA made investments, referred to collectively as NEO-WISE, to adapt existing software, add data pipelines, and to put in place data collection, archive and retrieval software for the new NEO-WISE database. Overall, NASA's investments in NEO-WISE resulted in the discovery of 146 previously unknown objects, including 129 near-Earth asteroids, of which 21 are Potentially Hazardous Objects, and 17 comets.

As noted in the following graphic, NASA continues to make progress toward the goals of the George E. Brown, Jr. Near-Earth Object Survey. To date, over 9,600 Near-Earth Asteroids of all sizes have been found. In addition to the 95 percent of NEOs over 1 km in size, our current estimate is that we have also found about 60 percent of the NEOs that are between 300 meters and 1 km.

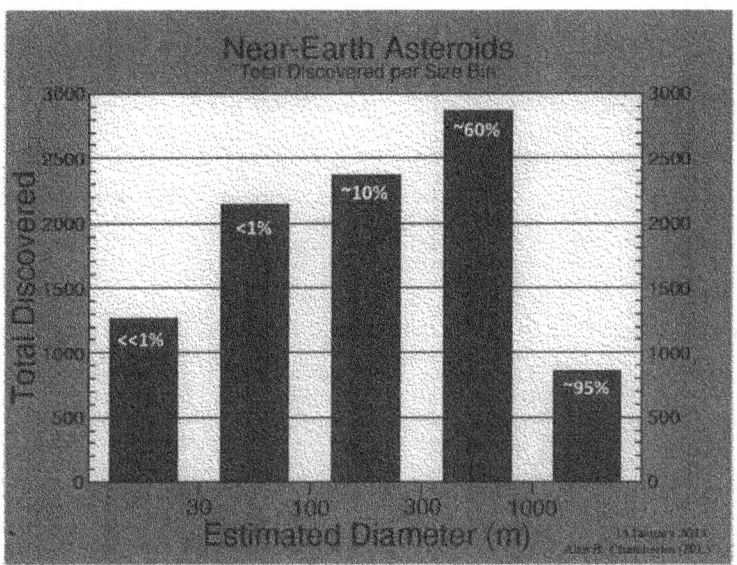

Risk Assessment and Characterization

Observations from all observatories around the world are sent to the international "clearinghouse" for small body observation data, the Minor Planet Center (MPC). The MPC maintains the database of observations and orbits on all known small bodies (asteroids, comets, dwarf planets, Kuiper Belt Objects, etc.) in the Solar System. It is funded by NASA and hosted by the Smithsonian Astrophysical Observatory's Center for Astrophysics. The MPC verifies and validates the observations by determining if they are of an already known object (by comparing them to the known orbits), or are indeed a new discovery. The MPC then determines and publishes an initial orbit for the new discovery so that observatories world-wide may look for the object and confirm its existence.

Once a new object's orbit is secured, its potential for impacting the Earth is assessed. As mentioned previously, well over 99 percent of all objects discovered have no potential for Earth impact even over many thousands of years, but the small fraction which do have some potential are tagged as PHOs by the Minor Planet Center. More detailed and refined analyses of a PHO's orbit, and an assessment of the risk posed by a particular object, are conducted by NASA's NEO Program Office at the Jet Propulsion Laboratory (JPL). Observations on PHOs are automatically forwarded to JPL and their orbits updated with high-precision analysis to determine a level of probability of the object impacting the Earth in the next 100 to 200 years. The results of this analysis are routinely updated and published on the NEO Program website at http://neo.jpl.nasa.gov.

When an NEO passes close enough to the Earth to be scanned by ground-based radar, NASA funds targeted radar observations. When an object passes close enough to the Earth to achieve a measurable radar return (about 20 million miles depending on the size), this allows NASA to obtain additional information about these objects. As was the case with 2012 DA14, the primary facility currently being used by NASA for routine planetary radar is NASA's own Goldstone facility, part of our Deep Space Network (DSN). NASA also uses the National Science Foundation's (NSF) Arecibo Radio Telescope facility for planetary radar observations. There are significant differences with the planetary radar capability at Arecibo compared to Goldstone. The Goldstone radar is a 70-meter steerable dish, allowing it to access objects significantly lower to the horizon than the more limited sky area accessible to the Arecibo radar. However, Arecibo is twice as powerful as Goldstone and has a much larger (304 meter) collection dish, which allows it to observe objects significantly farther away than Goldstone. Each plays an important role in the quick refinement of the orbit to a precision not obtainable by other means, and for understanding the object's size, shape and rotation rate. They can also aid in the detection of possible binary objects (~15 percent of NEOs), which in turn provides data that can be used to determine their mass.

NASA's NEO Observation (NEOO) Program has initiated development of several additional capabilities to the NEO detection network, with the recent additional funding it received starting in FY2012. Some of these involve collaboration on projects with the Defense Advanced Research Projects Agency (DARPA) and the U.S. Air Force, such as background detection of asteroids by the new Space Surveillance Telescope (SST), which is on track to start routinely providing observations this year. There is also the planned augmentation of the Panoramic Survey Telescope & Rapid Response System or Pan-STARRS facility with a second aperture. The wide field of view survey capabilities of these two assets are expected to provide a significant increase in NEO detection rate.

An important new development project that was started last year is the Asteroid Terrestrial-impact Last Alert System Project, or ATLAS Project. NASA's NEOO program has funded a five-year grant to the University of Hawaii to develop this innovative system. It will couple modest-sized, commercially-available telescopes with custom charge-coupled device cameras and rapid sky survey software to cover the entire available sky each night, detecting any asteroid bright enough to be seen by its detectors. ATLAS will find 100-meter sized asteroids millions of miles away, and smaller objects as they come closer to the Earth. It could provide days to weeks of warning of an object tens of meters in size that is on an impact trajectory with Earth. The prototype system is planned to begin sky testing by the end of calendar year 2014.

However, ATLAS, like SST and Pan-STARRS, will still be limited to the night sky and by weather. The only way to overcome these impediments is to use the vantage point of space. This is the idea behind a privately funded effort by the B612 Foundation to build a space observatory called Sentinel. NASA is providing B612 technical and operations assistance through a Space Act Agreement. Sentinel is being designed to find 100-meter sized objects and larger that could come near Earth's orbit.

To find the more numerous smaller asteroids near Earth, NASA also is investigating development of an instrument that could be hosted on geo-synchronous platforms such as communications, TV broadcast or weather satellites. This instrument would be a modest-sized, wide field telescope with detectors that operate in the infrared bands where these faint asteroids are more easily detected against the cold background of space. Though limited by the telescope size that can be hosted by a commercial geo-satellite, such a capability shows promise to increase the detection rate of near Earth asteroids. We plan to initiate this project with an instrument solicitation later this year. The NASA Science Mission Directorate is testing detectors initiated in its Discovery Program that may be used in such a hosted asteroid detection telescope.

And, to further leverage international capabilities, NASA has been a leading participant in the NEO activities of the United Nations Committee on the Peaceful Uses of Outer Space (UNCOPUOS). Over the past several years, a working group on NEOs under the UNCOPUOS Scientific and Technical Subcommittee has been examining the topic of Earth-threatening NEOs. Results of that work led to recommendations this year, endorsed by the Subcommittee, to broaden and strengthen the international network to detect and characterize NEOs, and to call for relevant national space agencies to form a group focused on designing reference missions for a NEO deflection campaign. NASA has been at the forefront of these activities and will continue to take on that role.

Reconnaissance Activities
NASA's remote ground-based observations of comets and asteroids have been augmented by close-up reconnaissance data from the agency's science missions. Since the main asteroid belt located between the orbits of Mars and Jupiter is the source of NEOs, it is important to study it to fully understand these objects. NASA's Galileo spacecraft flew by and studied the main-belt asteroids Gaspra in 1991 and Ida in 1993 on its way to Jupiter. NASA's Near Earth Asteroid Rendezvous or NEAR mission (later renamed NEAR Shoemaker) flew by the main-belt asteroid 253 Mathilde in 1997, and beginning in 1988 flew by, then orbited, and in 2001 landed on the near-Earth asteroid 433 Eros. NASA is cooperating with the European Space Agency on its Rosetta mission, which encountered the main-belt asteroids 2867 Šteins in 2008 and 21 Lutetia in 2010, and is scheduled to rendezvous and land on the Near-Earth comet 67P/Churyumov–Gerasimenko next year. NASA's Deep Space-1 spacecraft flew by a small Mars crossing asteroid, 9969 Braille, in 1999. NASA's Stardust spacecraft flew by the main-belt asteroid 5535 Annefrank in 2002. NASA's Deep Impact mission was specifically designed to impact and observe the effects on a comet. After imaging the comet Temple 1 in July 2005, the larger "flyby" spacecraft pointed high-precision tracking telescopes at the comet and released the "impactor" spacecraft into the comet's path for a planned collision. NASA's Stardust spacecraft later flew by Tempel 1 in February 2011, to further study what happened after the impact. NASA collaborated on Japan's Hayabusa mission which in 2010 successfully returned a small amount of samples from its 2006 encounter with the near-Earth asteroid 25143 Itokawa; 10 percent of these microscopic samples will be available for research by U.S. scientists as a result of NASA's agreement with the Japan Aerospace Exploration Agency (JAXA).

Most recently, NASA's Dawn spacecraft spent a year orbiting and observing the large main-belt asteroid 4 Vesta. Dawn provided close-up views of Vesta and unprecedented detail about the giant asteroid. Dawn's observations confirmed that the asteroid had completely melted in the past, forming a layered body with an iron core. The spacecraft also revealed the collisions Vesta suffered in its southern hemisphere. The asteroid survived two colossal impacts in the last 2 billion years. Without Dawn, scientists would not have known about the dramatic troughs sculpted around Vesta, which are ripples from the two south polar impacts. Dawn departed Vesta last August, and is now on its way to its planned rendezvous in 2015 with Ceres, the only dwarf planet in the inner solar system and the largest asteroid, so large that it is estimated to contain a third of the mass of the entire main asteroid belt.

These reconnaissance missions have greatly redefined what we know about asteroids and comets. These missions have observed asteroids that are binary systems, like Ida and Dactyl, where two objects travel together through space orbiting each other, which provides us insight into how these objects are formed and provide challenges for getting close to them. They have visited asteroids with primitive compositions that teach us about the origin of our solar system, and others that have undergone many of the same processes as terrestrial planets, providing a mirror for understanding our own. For example, data from our NEAR/Shoemaker mission suggests that Eros is a cracked but solid rock, probably a fractured chip off a larger body, made of some of the most primitive materials in the solar system. Judging from the meteorites, we have many more interesting varieties to visit in the future, from asteroids made of solid iron nickel alloys, to cosmic rubble piles, having collected bits and pieces over time and are only loosely connected.

A significant part of NASA's exploration of asteroids is to characterize the variety of asteroid composition and other characteristics that have both fundamental science value and provide critical information to inform the potential risk of PHOs. To this end, NASA is moving toward confirmation later this year for the Agency's planned asteroid rendezvous and sample return mission, dubbed OSIRIS-REx (for Origins-Spectral Interpretation-Resource Identification-Security-Regolith Explorer) that is planned to launch in 2016. After traveling three years, OSIRIS-REx will approach a near Earth asteroid, currently named 1999 RQ36, in 2019. 1999 RQ36 is believed to be the most exciting, accessible, volatile and organic-rich remnant currently known from the early Solar System; it also is thought to have abundant regolith (a blanket of loose materials covering rock), comprised of fine gravel that is ideal for collecting a sizable sample. With this sample, scientists will be able to analyze the asteroid's composition, mineralogy and geology to learn more about it and other organic-rich B-type asteroids. Once within three miles of the asteroid, the spacecraft will begin six months of comprehensive surface mapping. The science team then will pick a location from where the spacecraft's arm will take a sample of between 60 and 1000 grams (up to 2.2 lbs) for return to Earth in 2023. The careful study of 1999 RQ36 will permit scientists to fully understand the context in which the sample was selected, which should greatly increase the scientific value of the sample. NASA recently sponsored a contest for students worldwide to re-name 1999 RQ36 and draw attention to the important issues surrounding NEOs. The contest deadline was December 2, 2012, and we anticipate announcing a winner in the coming months.

NASA is also in discussions with our international partners to collaborate on several missions or mission concepts that could, in the future, grant access to U.S. researchers to valuable data on asteroids. NASA is working with JAXA on potential collaboration on the Japanese-led Hayabusa II mission. NASA is also discussing with the European Space Agency potential collaboration on two of their mission concepts: 1) the Marco-Polo-R mission concept which is focused on returning a sample from a primitive near-Earth asteroid in the late 2020s, and, 2) the Asteroid Impact and Deflection Assessment (AIDA) mission concept that could be used to study the **binary asteroid system Didymos with two spacecraft and see if a small interceptor can affect any the change in the relative orbit of the two bodies.**

Finally, NASA is working to accomplish an astronaut visit to an asteroid by 2025. This mission, and the vital precursor activities that will be necessary to ensure its success, should result in additional insight into the nature and composition of NEOs and will increase our capability to approach and interact with asteroids.

Conclusion
NASA has a long history of observing comets and asteroids but as their importance has become apparent as potentially hazardous objects, NASA has significantly increased its program of detection, reconnaissance, and characterization. NASA's current program utilizes extensive ground-based telescope observations in partnership with academia, U.S. Air Force, the National Science Foundation, and many international groups as well. We have a nearly complete understanding of our largest NEO population. NASA has determined that it is unlikely that the world will suffer a global catastrophic impact over the next several hundred years similar to the dinosaur extinction event. We are making marked progress in assessing the risk to our planet from smaller objects that could produce regional disasters. NASA is regularly reevaluating the risk to our planet and constantly updating our knowledge of the NEO population. Smaller objects, such as the recent impact in Russia, will always be difficult to detect and provide adequate warning; however, progress will be made over the next decade in this area as well.

Again, thank you for the opportunity to testify today, and I look forward to responding to any questions you may have.

National Aeronautics and Space Administration

Charles F. Bolden, Jr.

Nominated by President Barack Obama and confirmed by the U.S. Senate, retired Marine Corps Maj. Gen. Charles Frank Bolden, Jr., began his duties as the twelfth Administrator of the National Aeronautics and Space Administration on July 17, 2009. As Administrator, he leads the NASA team and manages its resources to advance the agency's missions and goals.

Bolden's confirmation marks the beginning of his second stint with the nation's space agency. His 34-year career with the Marine Corps included 14 years as a member of NASA's Astronaut Office. After joining the office in 1980, he traveled to orbit four times aboard the space shuttle between 1986 and 1994, commanding two of the missions. His flights included deployment of the Hubble Space Telescope and the first joint U.S.-Russian shuttle mission, which featured a cosmonaut as a member of his crew. Prior to Bolden's nomination for the NASA Administrator's job, he was employed as the Chief Executive Officer of JACKandPANTHER LLC, a small business enterprise providing leadership, military and aerospace consulting, and motivational speaking.

A resident of Houston, Bolden was born Aug. 19, 1946, in Columbia, S.C. He graduated from C. A. Johnson High School in 1964 and received an appointment to the U.S. Naval Academy. Bolden earned a bachelor of science degree in electrical science in 1968 and was commissioned as a second lieutenant in the Marine Corps. After completing flight training in 1970, he became a naval aviator. Bolden flew more than 100 combat missions in North and South Vietnam, Laos, and Cambodia, while stationed in Namphong, Thailand, from 1972-1973.

After returning to the U.S., Bolden served in a variety of positions in the Marine Corps in California and earned a master of science degree in systems management from the University of Southern California in 1977. Following graduation, he was assigned to the Naval Test Pilot School at Patuxent River, Md., and completed his training in 1979. While working at the Naval Air Test Center's Systems Engineering and Strike Aircraft Test Directorates, he tested a variety of ground attack aircraft until his selection as an astronaut candidate in 1980.

Bolden's NASA astronaut career included technical assignments as the Astronaut Office Safety Officer; Technical Assistant to the director of Flight Crew Operations; Special Assistant to the Director of the Johnson Space Center; Chief of the Safety Division at Johnson (overseeing safety efforts for the return to flight after the 1986 Challenger accident); lead astronaut for vehicle test and checkout at the Kennedy Space Center; and Assistant Deputy Administrator at NASA Headquarters. After his final space shuttle flight in 1994, he left the agency to return to active duty the operating forces in the Marine Corps as the Deputy Commandant of Midshipmen at the U.S. Naval Academy.

Bolden was assigned as the Deputy Commanding General of the 1st Marine Expeditionary Force in the Pacific in 1997. During the first half of 1998, he served as Commanding General of the 1st Marine Expeditionary Force Forward in support of Operation Desert Thunder in Kuwait. Bolden was promoted to his final rank of major general in July 1998 and named Deputy Commander of U.S. Forces in Japan. He later served as the Commanding General of the 3rd Marine Aircraft Wing at Marine Corps Air Station Miramar in San Diego, Calif., from 2000 until 2002, before retiring from the Marine Corps in 2003. Bolden's many military decorations include the Defense Superior Service Medal and the Distinguished Flying Cross. He was inducted into the U.S. Astronaut Hall of Fame in May 2006.

Bolden is married to the former Alexis (Jackie) Walker of Columbia, S.C. The couple has two children: Anthony Che, a lieutenant colonel in the Marine Corps who is married to the former Penelope McDougal of Sydney, Australia, and Kelly Michelle, a medical doctor now serving a fellowship in plastic surgery.

www.nasa.gov

Chairman SMITH. Thank you, Administrator Bolden.

I recognize myself for questions, and let me address the first one to Dr. Holdren and then perhaps, Administrator Bolden, to you as well. There seems to be general agreement based upon your testimonies that we are able to detect 90 to 95 percent of the near-Earth objects that are larger than 1 kilometer somewhere around 60 percent of the objects that are over 300 meters, so my question is this. I haven't heard yet nor have I seen yet what percentage of the near-Earth objects, the incoming asteroids that are 100 meters, what percentage of those objects are we able to detect, 100 meters being, I think, Dr. Holdren, you described in your written testimony as the size of a city destroyer. So what percentage of the 100-meter near-Earth objects can we detect, and do you have a figure for that?

Dr. HOLDREN. I believe at this point that number would be a little under 10 percent. The number for 140 meters and above is 10 percent. The 100 would be a little under 10 percent.

Chairman SMITH. Administrator Bolden, do you agree with that?

General BOLDEN. Yes, sir, that was on that second chart I showed where it looks like the less than 10 percent for——

Chairman SMITH. Okay. How many objects are we talking about that we are not able to detect that might be the city destroyers?

General BOLDEN. Numbers of objects?

Chairman SMITH. Yes.

General BOLDEN. Mr. Chairman, I don't know that answer, and that is one thing I cannot take for the record because——

Chairman SMITH. What was the 10 percent?

Dr. HOLDREN. I can answer that question, Mr. Chairman.

Chairman SMITH. Okay, Dr. Holdren.

Dr. HOLDREN. The estimates of how many objects exist, near-Earth objects in the range of 140 meters or above are between 13,000 and 20,000 objects. So that is the number of which we have detected 10 percent. That is the much more challenging goal, which the Congress put before us to identify 90 percent of those by 2020.

Chairman SMITH. Roughly 2,000 objects that are city destroyers, we are not detecting. Is that roughly right?

Dr. HOLDREN. No, more, because the number we are detecting is 10 percent of 13,000 to 20,000 so——

Chairman SMITH. I was going in——

Dr. HOLDREN. So you were going the other way. Unfortunately, the number undetected——

Chairman SMITH. I was going 1,300 to 2,000, and I was going to the larger figure. That is why I said 2,000.

General BOLDEN. So the number of undetected potential city killers is very large. It is in the range of 10,000 or more.

Chairman SMITH. Ten thousand or more. Okay. Not reassuring, but what is reassuring, we hope, is the unlikelihood that one of those city destroyers would actually hit a city. As you pointed out, two to three percent of the earth's area is urban area.

Administrator, what programs, what improvements, what developments can we expect in the next, say, 2 years or 5 years to be able to better detect these thousands of near-Earth objects that might be life threatening?

General BOLDEN. Mr. Chairman, we continue our work, our collaboration with our international partners. That is very important. As Dr. Holdren mentioned earlier, he didn't specify but it was a Spanish astronomer, amateur astronomer actually, or I think——

Chairman SMITH. Do you expect improvements in Earth-based—I mean telescopes, for example, that will enable us to better detect these?

General BOLDEN. What we are really looking at is not improvements but increase in the numbers of space-borne assets. We really need to have space-borne assets that are able to look. We are cooperating right now with a Space Act Agreement with a private company called B612 that will be engaged in the identification and characterization of asteroids, and my hope is that there will be more.

Chairman SMITH. Okay. And what percentage of these thousands would we be able to detect in the next few years that we are not detecting now? Any idea?

General BOLDEN. If you talk about the 140-meter class, our estimate right now is at the present budget levels—that is present budget levels, not the going-down budget levels—it will be 2030 before we are able to reach the 90 percent level as prescribed by Congress to detect and characterize those 90 percent of the 140-meter class.

Chairman SMITH. Okay. Thank you for the answer, though, again, that is not particularly reassuring. Maybe we can help you out with the budget. Don't know.

General Shelton, last question for you. Was the Department of Defense aware of the meteor that exploded over Russia?

General SHELTON. Mr. Chairman, not until we were tipped off by NASA.

Chairman SMITH. And that was after the fact, or how far before the fact?

General SHELTON. No, it was—I want to say it was two or three days preceding——

Chairman SMITH. Two or three days before it exploded over Russia? Okay.

General SHELTON. I am sorry. You said the explosion. I was talking about DA14.

Chairman SMITH. No, I am talking about the meteor that exploded over Russia.

General SHELTON. We had no insight in that at all.

Chairman SMITH. Even with satellites, even with everything else?

General SHELTON. We were aware of the event when it occurred.

Chairman SMITH. And not before?

General SHELTON. Not before.

Chairman SMITH. I just have to ask you, how then are we going to be aware of, say, incoming missiles if we couldn't detect the meteor exploding over Russia?

General SHELTON. Now, we did detect it. We were aware of the event.

Chairman SMITH. But at the time of the event, not before?

General SHELTON. Yes, sir, and we would have to take that into a different forum to talk in more detail.

Chairman SMITH. Okay. Thank you, and that concludes my questions. The Ranking Member, Ms. Johnson, is recognized for hers.

Ms. JOHNSON. Thank you very much.

Dr. Holdren, in October 2010, the Congressional response to the direction in the 2008 NASA Authorization Act described roles and responsibilities for NASA, FEMA, DOD and State but is silent who has the overall responsibility, and I was wondering who in this Administration is the—who has the single responsibility to oversee the other activities of other agencies?

Dr. HOLDREN. Well, NASA is responsible, has the overarching responsibility for detection and notification. NASA notifies FEMA, they notify the Department of Defense. On the question of mitigation, who would have the responsibility if an asteroid were discovered to be on a collision course, that would depend on the size of the asteroid and the amount of notice we had. For some deflection missions, you would want NASA to be in charge. For other kinds of deflection missions, you would want DOD to be in charge. So it does not make sense from the standpoint of the mitigation mission to specify in advance which agency would do it, but the notification—identification and notification responsibilities are unambiguous.

Ms. JOHNSON. So when there is mitigation, do all of you come together or who takes the lead? What determines who takes the lead?

Dr. HOLDREN. In that event, we would certainly all come together, and we are in fact exercising those kinds of communications. There is actually an exercise coming up in the middle of next month when we will exercise those interactions, communications and the exercise of responsibilities. There is a workshop actually coming up at the beginning of next month in which those interagency interactions will be further discussed and delineated.

Ms. JOHNSON. Thank you very much. Thank you, Mr. Chairman.

Chairman SMITH. Thank you, Ms. Johnson. The gentleman from California, Mr. Rohrabacher, is recognized for his questions.

Mr. ROHRABACHER. Thank you very much, Mr. Chairman.

We are talking about space debris and near-Earth objects that are—it seems to me that these two issues are not just American issues, and we are talking about the cost all this, what are we talking about in terms of over a 20-year period, the costs of actually coming up with a deflection and the cost of actually making the determination of what is heading in our direction? Dr. Holdren, or do any of you have estimates of cost?

General BOLDEN. Mr. Rohrabacher, I can give you an estimate right now. We do it incrementally so we believe we have to detect and characterize first and then we have to concern ourselves, as Dr, Holdren says, with who is going to do the mitigating action or the deflection action. We have two concepts. One is about three-quarters of a billion dollars for an infrared-based sensor that is placed in space, something that orbits Venus or at least is in geo-synchronous orbit. B612, that I mentioned, their estimate for their effort is about a half a billion dollars, about $500 million dollars. So we are roughly in that range.

Mr. ROHRABACHER. Is that just for that one sensor that we are talking about?

General BOLDEN. That is just for—to try to put something in space that will help us to identify and characterize. I think all three of us agree, ground-based systems are great, Arecibo and others, but if you really want to find and detect asteroids, near-Earth objects early enough that we can do something, then you want that vehicle——

Mr. ROHRABACHER. And the cost is?

General BOLDEN. I gave you an example of two. I will take it for the record to get back to you. I think what you are asking for is a lifecycle cost for a program to mitigate.

Mr. ROHRABACHER. Right.

General BOLDEN. I don't think any of us have—we have not developed that.

Mr. ROHRABACHER. Well, it is in the billions of dollars, correct?

General BOLDEN. Oh, yes, sir.

Mr. ROHRABACHER. Okay.

General BOLDEN. You know, if one detection device is almost a billion——

Mr. ROHRABACHER. Now, let me suggest that perhaps the billion dollars, and that would provide protection for not just the United States but for the world.

General BOLDEN. Sir, anything we are talking about—this is not—as you pointed out, this is not an American issue. Anything that we do protects the planet. Anything that our international partners do protects the planet, and that is why you hear me talk all the time about the critical importance of international collaboration.

Mr. ROHRABACHER. That is what I want to ask you about on this. What steps have we taken to bring countries together that could contribute those billions of dollars as well as our own?

General BOLDEN. Well, the U.N. Organization for Peaceful Cooperation of Space, U.N. COPUOS, has a very active ongoing activity and trying to help bring nations together and looking at detecting and tracking NEOs. We are a participant in that.

Mr. ROHRABACHER. There is not just one organization that is aimed specifically or—when was the last meeting of groups of people who represent countries that might want to get involved and contribute and have an overall part?

Dr. HOLDREN. Congressman Rohrabacher, I can take that one. There was a meeting in Vienna in mid-February of this year just a month ago under the auspices of the U.N. Committee on Peaceful Uses of Outer Space. It was agreed there to stand up an international asteroid warning network and to stand up as well an international body that would deal with the mitigation question. There is already underway something called AIDA, the Asteroid Impact and Deflection Assessment, which is a joint effort of the European Space Agency and NASA, and I should add that the detection network that we already have is highly international in character. As Administrator Bolden mentioned, it was actually a Spanish observer who first discovered the asteroid that made the near miss on February 15. The Minor Planet Center, which is in substantial part funded by NASA and hosted by the Harvard-Smithsonian Astrophysical Observatory, is actually under the overall auspices of the International Astronomical Union, so it is all very international.

Mr. ROHRABACHER. I would suggest that number one, the cost of deflection of course, we are talking about the cost of detection, in one situation, the cost of having a deflection system is even more. I would suggest that this is one area of leadership that the United States could really take a role in and it would be good for all and it would create an international spirit of what we want to create. I would suggest especially including Russia in on this, and they may be able to make some major contributions, save us some money and actually make it a more effective system.

And with that said, I would like to include all countries except China. Thank you.

Chairman SMITH. Thank you, Mr. Rohrabacher. The gentlewoman from Maryland, Ms. Edwards, is recognized.

Ms. EDWARDS. Thank you, Mr. Chairman.

I want to ask Dr. Holdren, the National Science Foundation has indicated a next major new start as the Large Synoptic Survey Telescope, the LSST, which is intended to detect and catalog potentially hazardous objects, and what I would like to know is, one, what the technological contribution would be if the LSST were to make the overall detection and cataloging effort possible, and General Bolden, you talked about the prospect of land-based systems versus systems that we would put outside in our solar system, but the cost, to me, it seems would be rather significantly different. And then I would like to have some understanding of whether there might be some cost sharing that NASA might consider with improvements to the LSST to try to optimize it for NASA's use, and get a sense as well of whether the challenges that we are facing and not meeting the 2025 deadline that—guideline that we have highlighted from the Committee. Are those technological challenges principally? Are they funding challenges? Is it some combination of cooperation challenges? I would like to better understand that, especially in this fiscal environment.

Dr. HOLDREN. Well, let me just make a start and then I will turn it over to Administrator Bolden. The Large Synoptic Survey Telescope would be an important addition to our capabilities but it is important to understand that all these capabilities work in tandem, that is, they share information. Some of the telescopes are better at detection. Others are better at characterizing the orbit or determining the reflectivity and the likely composition of the object, and so one always has to think of this as a network. We have telescopes in Arizona, we have telescopes in Italy, we have telescopes in the Czech Republic, and they are all linked together and they are all part of a network that provides the overall capability we have to detect these objects. The LSST alone when it comes fully to fruition would still not be able to enable us to identify and characterize 90-plus percent of the objects in less than about a dozen years. But in combination, the LSST and an orbiting infrared telescope of the kind Administrator Bolden was talking about could lower that time to something in the range of 6 to 8 years.

General BOLDEN. Congresswoman, the only thing I will add, you know, we flew an infrared imaging satellite called WISE, and then we repurposed it while on orbit to look for asteroids, and we discovered hundreds in the deep field of the solar system, the universe, actually. It is that type of instrument that I talk about. That is what B612 wants to do. We are looking at ways to cost-share. The nucleus organization that Congressman Rohrabacher mentioned involving Russia, the 5-member organizations of what we call the International Space Station team, and that is 15-plus European nations, Russia, Japan, Canada and the United States, although our primary responsibility is operating the International Space Station, when the heads of agency get together, we talk about everything, and one of the big things we talk about is the threat of near-Earth asteroids.

At risk of getting in trouble because Congressman Rohrabacher and I have a healthy agreement to disagree, and I will say this, it will be the decision of this Congress as to whether or not we ever cooperate or participate with China. It is the elephant in the room. I don't talk about it because my public affairs and communications people tell me not to talk about it, but I don't deal with China by direction of this Congress. We are the only agency of the Federal Government that does not have bilateral communications with China. This is an issue for the world. This is not an issue for the United States, so although Congressman Rohrabacher and I——

Ms. EDWARDS. Well, I will let Congressman Rohrabacher take his time talking about China, and I am sure we could have a whole hearing on it. Before we go, though, I wanted General Bolden to— you know, the whole identified mission that the President has set out to go to an asteroid, it seems rather lackluster, and so I have always had questions about whether ought to be a goal or we ought to think about, you know, sort of the tradeoff, Mars, instead. Thank you.

Chairman SMITH. Thank you, Ms. Edwards. The gentleman from Texas, Mr. Hall, chairman emeritus, is recognized for his questions.

Mr. HALL. Mr. Chairman, of course I thank you for holding this very important hearing, and I thank the witnesses for their very valuable testimony.

I had the privilege of serving on this Committee since 1981, and this topic has been the subject of periodic review and legislative direction, as the witnesses noted, in the 1990s during consideration of a NASA authorization bill. This matter came up, and it was really a discussion about asteroids. We had really a hearing on asteroids, as Mr. Rohrabacher remembers, and it was reported at that time that one had just passed the Earth that no one knew anything about but it missed us by 15 minutes. I hated to ask, was that just as good as it missing us by 1 minute or 30 seconds or what, but just the enormity of the damage that they could do to us. I offered an amendment at that time to set a goal of finding and cataloging within 10 years this population of comets and asteroids in an effort to be coordinated with the Department of Defense and space agencies of other countries. Other countries were invited to that hearing, but also told that we ought to have a world group because as Charlie said, it is a world problem. They were interested in attending but they weren't interested in contributing anything to it, so none of them showed up for the hearing.

But as our witnesses stated, from 1998 until 2011, more than 90 percent of near-Earth objects with a diameter of 1 kilometer or greater have been located. So today we know more about these but we also have more work to do, especially those that are smaller that could still have a devastating impact if they hit the Earth.

So Dr. Shelton, let me ask you this. What capabilities do we need that we don't currently possess to detect and track asteroids that might pose a threat to the Earth?

General SHELTON. Sir, if you are talking about Department of Defense capabilities——

Mr. HALL. What do we have to do? What should we do?

General SHELTON. Well, if you are talking about Department of Defense capabilities, we are focused on things in Earth orbit. Our sensors, and we have got a variety of them, are not focused on beyond the Earth.

Mr. HALL. Well, once an object has been identified, what are our means of tracking it and how much time would we have to prepare if there were a threat to Earth?

Dr. HOLDREN. Maybe I can take that, Congressman Hall. First of all, how much notice we have depends on the size of the object. The bigger it is, the further away we can see it and the more time we have. So there are some objects that we know are coming years in advance. There are other objects that are still big enough to cause damage that we only know about weeks in advance or days in advance. Obviously, we need to improve the capability to give us a large amount of notice, enough notice to mount a deflection mission if we see one on a collision course. Some of the capabilities we have been talking about, the Large Synoptic Survey Telescope, the orbiting telescope that the B612 Foundation is working with NASA to develop, all those capabilities will increase the warning time with respect to asteroids big enough to do serious damage. And again, the deflection options that would then be open to us would depend on the size of the object and the amount of notice we had. They would include——

Mr. HALL. Well, excuse me. The one that hit Russia, there is no question about that, and that is about all we know about it, why didn't we know that was coming or on its way?

Dr. HOLDREN. It came out of the sun, Congressman Hall. It came from a direction where our telescopes could not look. We cannot look into the sun.

Mr. HALL. Well, if we can't make that determination as to where it is going to come from, we ought to be able to do something no matter where it comes from if it is going to hit the Earth.

Dr. HOLDREN. That is one of the reasons that an orbiting telescope——

Mr. HALL. That is why we are having this hearing today to ask you three men who know a heck of a lot more than we know about it to tell us.

Dr. HOLDREN. Well, I would say, Congressman Hall, that the most important single thing we could do to improve our capacity to see any asteroid of potentially damaging size coming would be an orbiting infrared telescope of the sort that the B612 Foundation is working on.

Mr. HALL. I thank you. I asked the question, if we saw one come toward Omaha, what could they do about it, and they said they could use a laser, and I went on and asked a second question. I said, well, could the laser hit it right in the middle because I didn't want to cause any more trouble than I had with Mr. Rohrabacher. I wasn't going to suggest that half of it hit Los Angeles and the other half hit New York. I suggested that half of it might go to the Pacific Ocean and the other half go to the Atlantic Ocean. They really didn't have an answer for that, and I doubt if you have.

Dr. HOLDREN. Well, first of all, it would not be practical to have a laser powerful enough to split it in half. What you can do in principle if you have a very powerful laser is to cause jets of material heated by the laser to fly off of the asteroid and that is essentially the equivalent of a jet engine pushing the asteroid off course. There are other approaches to deflecting an asteroid. Those include hitting it with a very heavy impacter. They include approaching it, as we have already approached with robotic probes a number of asteroids and pushing it or towing it.

Mr. HALL. I thank you, and I will write you a letter for some more, and thank you. I yield back my time.

Chairman SMITH. Thank you, Mr. Hall. Those were interesting answers, Dr. Holdren. I appreciate that.

The gentlewoman from Oregon, Ms. Bonamici.

Ms. BONAMICI. Thank you very much, Mr. Chairman, and thank you all for your interesting testimony.

It has been well established in this testimony that the probability of an occurrence of a sizable NEO colliding with the Earth is quite small. I believe, General Bolden, you said extremely remote in your testimony. But it is also clear that the consequences could be enormous. For example, a strike, depending on the size of an asteroid, could bring a cloud of dust rivaling the most powerful volcanic explosion, or depending on where it hits could cause an enormous tsunami that would flood and destroy coastal regions. And I know you are all striving as we are to find the appropriate balance for investment without being unnecessarily alarmist.

In the district—back to where it hits. In the district I represent in Oregon, there is a significant threat of a tsunami, especially from earthquakes. That is very real. Response preparedness is already a priority issue for my constituents. In fact, when I was in the legislature, we passed a bill that required the State to plan for the impacts of a 9.0-magnitude earthquake and a resulting tsunami, which scientists had determined would occur, will occur at some point in the future, so it is not planning for if, it is planning for when. And the State just released its resilience plan, which was partially funded through a FEMA grant, in February. The plan acknowledges the importance of preparing communities and infrastructure for a catastrophic event but it also places significant focus on the ability to respond once the event has occurred.

And of course, this type of challenge has implications in the context of today's conversation. How much do we plan for detection, how much do we plan for response? Of course, we should be investing in the science that will help us detect and prevent the impacts of NEOs but we also need to consider how we will respond if it not possible to alter the orbits and stop these NEOs from colliding.

Dr. Holdren, your 2010 report indicates that depending on the projected damage and location, FEMA could help provide Federal assistance and coordinate local emergency services personnel into integrated disaster response task forces. So could you talk a little bit more, please about how FEMA is approaching this role? How will they take into account different demographic and geographic characteristics in any given area? Thank you.

Dr. HOLDREN. Wow, that is a really challenging question. You know, as we know, FEMA has a wide range of capabilities for responding to a wide variety of different kinds of emergencies and disasters. We are in the process, as I mentioned, of conducting exercises of various kinds in which FEMA is a participant, and thinking about and trying to work out the details of response strategies, depending on the nature of the impact, but as your question points out, those impacts could be very different. If a large asteroid strikes the ocean, as you point out, the impacts would largely come through the tsunami phenomenon, which is of course a phenomenon with which FEMA must also reckon since tsunamis can be caused in other ways. If a strike occurred over an urban region with sufficient force, the damage would resemble in some ways the damage from a massive earthquake, which is another event with which FEMA is familiar and prepared to respond. But these are going to be big challenges. I would not minimize the difficulty of responding adequately if a substantial asteroid strike should occur in the size range that we need to be particularly worried about.

Ms. BONAMICI. And so what efforts are being made to engage the existing emergency response infrastructure?

Dr. HOLDREN. Well, as I say, we are actually exercising those with tabletop exercises and with larger-scale exercises in which the various agencies go through a simulated event of this kind, and those kinds of exercises are really the best way we have when combined with analytical tools to figure out how to bring our capabilities effectively to bear.

Ms. BONAMICI. Thank you very much.

And either General Bolden or General Shelton, do you have any comments about finding that balance between preparing for detection and preparing for how we respond?

General BOLDEN. Congresswoman, I would just echo what you said. You hit the right word, and that is balance. You know, we could come out of this hearing and decide that we want to really pour money into NEO detection and characterization, and that would not be the right thing to do because there has to be a balance. My recommendation would be the President's budget from 2013, I think was pretty good. We have a plan that Dr. Holdren talked about but it depends on the passage of that budget. Going into 2014, we will come back again and try to give you what we see as a funding level to support a plan that Dr. Holdren addresses. So that is where we have to cooperate, Congress and the Administration, in striking that proper balance.

Ms. BONAMICI. Thank you very much. My time is expired. I yield back. Thank you, Mr. Chairman.

Chairman SMITH. Thank you, Ms. Bonamici. The gentleman from Alabama, Mr. Brooks, is recognized.

Mr. BROOKS. Thank you, Mr. Chairman.

Reading from Dr. Holdren's testimony, it says "Depending on its composition and velocity, an asteroid of 140 meters in diameter could have an impact energy in the range of 50 to 500 megatons of TNT equivalent and would be capable of causing destruction over a large region,'' emphasis there 50 to 500 megatons, and I have got other notes here that suggest that the Hiroshima atomic bomb was roughly 13 kilotons, so much, much, much, much smaller. If you could, could you please describe with greater detail what you mean by a ''large region''?

Dr. HOLDREN. The size you are talking about, 140 meters, and you have got the numbers exactly right, could devastate the better part of a continent.

Mr. BROOKS. We are talking about a very large region.

Dr. HOLDREN. The fortunate—the only fortunate thing is that the estimated frequency with which objects of that size strike the Earth is about one in 20,000 years, or a probability of one in 20,000 each year. Nonetheless, this falls directly in the category that we were talking about, low probability, very high consequence, therefore we need to take the risk seriously and we need to make the kinds of investments that would enable us to deflect an asteroid of that size were one to be discovered on a collision course.

Mr. BROOKS. And you also used the word ''destruction'' in the context of this continent-sized area. Would human life be able to withstand that kind of impact and the way in which you use the word ''destruction''?

Dr. HOLDREN. Well, clearly, if an asteroid of that size struck on land, there would be very large loss of life. If it struck in the ocean, it would produce, in all likelihood, a very large tsunami, which would be associated with large loss of life. If you say would humans survive on the Earth, the likelihood is yes. But there are concerns about the amount of dust and smoke that could be lofted into the atmosphere by such an impact.

Mr. BROOKS. Do you have a judgment as to whether humans would survive on the continent impacted, if you limit it just to the impact continent?

Dr. HOLDREN. No, I believe the answer is yes. Is aid a substantial part of a continent. A bigger one, bigger still than 140 meters, could be a continent destroyer, and a bigger one still could be a civilization destroyer. You know, the one that hit 65 million years ago near what is now the Yucatan Peninsula is thought to have led to the extinction of the dinosaurs and most else that lived on Earth at the time.

Mr. BROOKS. And if I read your written testimony correctly, that was roughly 10 kilometers estimated size?

Dr. HOLDREN. Yes.

Mr. BROOKS. Moving on, looking at the notes that I have been given by the HASC Committee, it suggests that we have identified so far thousands of objects in space, near-Earth objects in space, that are 300 to 500 meters in diameter, roughly 1,100, 1,200 that are roughly 500 to 1,000 meters in diameter, and roughly 900 that are a kilometer or more in diameter. So what I would like to know is, how much advance warning would the Earth's population need if, say, one of these kilometer or larger size objects for us to be able to do something to prevent that object from hitting the Earth and causing the kind of massive devastating that you have described?

Dr. HOLDREN. Today, we would probably need years to mount such a mission. Over time, as we develop our capabilities to deal with this kind of threat, the lead time could be smaller.

Mr. BROOKS. Let me focus in on that. How many years would we need? Let us say we found out today that there is an object of this size that is going to hit the Earth. How many years would we need today if we were to do whatever is necessary to try to put ourselves in a position to save the planet?

Dr. HOLDREN. I think I will refer that question to General Bolden.

General BOLDEN. Well, if we did it according to the President's budget presently, 2025 is the time that we think we will be able to send a human to an asteroid acting with some robotic means. That is on——

Mr. BROOKS. Let me interject for a moment. Let us assume that we know one is going to hit the planet, in which case I assume that we are going to accelerate things as quickly as we can. What is the fastest we can get it done where we could protect ourselves upon discovery of a 1-kilometer or larger object going to hit the Earth?

General BOLDEN. Congressman, I will take it for the record and get back to you, but now you are talking about an intense effort, which, I mean, that significantly shortens the time.

Mr. BROOKS. Well, we would be intense.

General BOLDEN. We have the systems and the technology available now to do that. You are talking about just pouring unlimited funds into it, and conceivably you could do it in 4 or 5 years. I don't know. But let me get back to you. Don't quote me on a number yet. But, I will work with General Shelton and his captain and, seriously, we will get you an answer.

Mr. BROOKS. Well, thank you for being here and testifying before us. Thank you, Mr. Chairman, for the time that you have allotted, and whatever time that is, I would love to help you shortening it.

Chairman SMITH. Thank you, Mr. Brooks. The gentleman from California, Mr. Swalwell, is recognized for his questions.

Mr. SWALWELL. Thank you, Mr. Chairman, and thank you, Ranking Member Johnson.

General Bolden, I represent Livermore, California, which has two of the NNSA labs, Lawrence Livermore and Sandia, and I imagine that when you talk about systems and technology, and if we were to require a weapon to deflect something that was incoming, a near-Earth object that was incoming, that some of that technology will have to be or has been designed at one of those laboratories.

General BOLDEN. So if that a question——

Mr. SWALWELL. Yes.

General BOLDEN. If that were the decision, but again, I would go back to what Dr. Holdren said earlier. I would not consider a weapon to deflect or to save Earth against this type of threat. I would consider the development of appropriate technologies that could enable us to—we are talking about earliest detection, you are talking about deflecting. I mean, it is a tiny amount if you catch it far enough out.

Mr. SWALWELL. Let us assume that it is late-stage detection. I imagine our choices get limited, right?

General BOLDEN. Yes, sir. That is not my bailiwick anymore. I don't do bombs and rockets.

Mr. SWALWELL. Well, General Shelton, those two laboratories in my district, I imagine they would play a critical role if we had a late-stage detection of one of these near-Earth objects.

General SHELTON. Yes, sir, I would think so. I mean, there are only a limited number of ways to generate the amount of energy required and probably nuclear energy is what we are talking about here.

Mr. SWALWELL. Is there a way to guarantee that one of these near-Earth objects does not hit on a Friday? Because right now in my district, all of the Federal employees at those laboratories are furloughed on Fridays. And I know in Congresswoman Edwards' district, some of those NASA employees that are trying to detect these incoming objects, I think they are going to be furloughed on Fridays too. So——

General BOLDEN. No, sir.

Mr. SWALWELL. No way to——

General BOLDEN. We are not planning to furlough employees. I just wanted to clarify that. So they will be there on Friday.

Mr. SWALWELL. Okay.

General BOLDEN. But in seriousness, I have to go back again to say several things. One, these are remote occurrences. Two, the plan that the President has put forward I think will adequately address our technical capability to be able to deflect an asteroid in due time. If we find that we are tracking literally thousands of asteroids today. If the civilization destroyer that Dr. Holdren talks about, I mean, if we can't discover that early enough, then there is something wrong with our systems.

Mr. SWALWELL. Sure. So in our district, it is a fact: there are furloughs at our nuclear laboratories, and you are not concerned at all that sequestration affects our readiness to protect——

General BOLDEN. Sir, that wasn't the question you asked.

Mr. SWALWELL. So my question is——

General BOLDEN. I am very concerned with the effects of sequestration but that wasn't the question, and so yes, I am very concerned about the effects of sequestration on all of our ability to do what it is you ask us to do. You are talking about impacting our ability to keep our facilities operating safely. You are talking about just the mental strain on our employees not knowing whether they are going to be able to come to work tomorrow. I try to assure them every time I can that I am not planning to furlough anybody, but they know better than I do that the Congress could take some action and all of a sudden the Administrator doesn't have a clue what he is talking about because now I have got to lay people off. My intention is not to do that. If your question is, is there a bad effect of sequestration, yes, sir.

Mr. SWALWELL. That is my question.

General BOLDEN. Yes, sir.

Mr. SWALWELL. How about for General Shelton?

General SHELTON. I will tell you, sir, just about my every waking moment these days is based on this topic. I just pulled the trigger on $508 million of reductions in just my major command alone from now until the end of the fiscal year, a 20 percent cut in pay to my civilians. There are resources that are used for missile warning and missile defense that we won't be able to operate at full capability. There are things that we use for space surveillance that we won't be able to operate at full capability.

Mr. SWALWELL. And General, do you think that makes us more or less prepared to handle a near-Earth objects?

General SHELTON. That is not what we do. That is NASA's responsibility. We contribute serendipitously at times but we are focused on things in Earth orbit.

Mr. SWALWELL. So if you had to focus on something in Earth orbit, would it make you more or less prepared having to have these across-the-board cuts?

General SHELTON. We are clearly less capable under sequestration.

Mr. SWALWELL. Great. Thank you, Mr. Chairman. I yield back the balance of my time.

Chairman SMITH. Thank you, Mr. Swalwell. The gentleman from Florida, Mr. Posey, is recognized for his questions.

Mr. POSEY. Thank you very much, Mr. Chairman, and thank all three of you for your very detailed written testimony. You use a lot of facts that I frequently refer to that clearly indicate it is not a matter of if but when civilization will be threatened by an impact. Until the recent Russian impact, quite a few people thought those of us who were even aware of this or dared mention it were on the kooky side, and so one good thing about that is maybe a little bit of a wake-up call to reality for some people.

Dr. Holdren, your testimony referred to the first-ever exercise, deflection exercise. I wonder if you could just share a little bit with us about how that went.

Dr. HOLDREN. I am not—the first-ever deflection exercise was a kinetic impact on an asteroid of medium size, which while interesting from the standpoint of the deflection it generated did not reflect the magnitude of the capability you would need for a late-notice deflection of an asteroid of threatening size. It was an interesting demonstration.

One of the things I would like to reinforce is that the President's proposal to land U.S. astronauts on an asteroid by 2025 will in fact exercise a number of the capabilities that would be necessary to have in our toolbox should an asteroid of threatening size be detected on a collision course. I would disagree with something Congresswoman Edwards said, that this is a lackadaisical program. I think it is a crucial program, and I think it is going to lead to major advances in capabilities which are not just interesting to demonstrate at a small scale but not enough to deal with a real threat.

Mr. POSEY. Thank you. And I took her comment to mean she thought the approach to it might have been lackadaisical, not that it wasn't necessary, you know, for whatever——

Ms. EDWARDS. For the record, I didn't say that word.

Mr. POSEY. Okay. Now, the Ranking Member asked about protocol, you know, who is in charge, and we got about three or four minutes of a chatter but we never got an answer about who is in charge, and so rather than asking for a response, I would just like to recommend that the next time that you all come before us you give us a protocol and say this is who is in charge here, here is in charge here and here is in charge here, and it is just a very clear matter of protocol who is in charge in various instances, you know, as being preordained and preestablished.. I know you are going to corroborate and, you know, get this stuff done if we have an impact, but a good segment of the population thinks it is just a matter of calling Bruce Willis in, you know, and notwithstanding we don't have a shuttle anymore, you know, it is impossible. But things that beg for an answer, you know, scary of course, that we only know about 10 percent of the huge threats and we virtually have no idea of the small threats like the one that went undetected, the recent impact in Russia. You know, what would we do if you detected even a small one like the one in Russia headed for New York City in three weeks? What would we do? Bend over and what?

General BOLDEN. No, Congressman, I have to go back to what I said before. These are very rare events. From the information that we have on asteroids that we have discovered of all sizes, we don't know of one that will threaten the population of the United States in three weeks, and we are trying very diligently as I said before with the President's budget to put ourselves in a position where we advance the technologies so that three weeks will not be something that causes us to panic because we will be able to respond.

We are where we are today because you all told us to do something, and between the Administration and the Congress, the bottom line is always the funding did not come, and I don't care whose fault it is or if it is anybody's fault. We all know what we are facing today and we are all sitting here today as the Congress and the Administration try to figure out sequestration, something that never should have happened. Nobody planned it to happen but we are facing it today. And so the answer to you is, if it is coming in three weeks, pray, if we find that out right now. And that is not bad policy.

Mr. POSEY. That is reality.

General BOLDEN. I am a practicing Episcopalian and I love what the Pope is doing right now. I will tell you, things have happened. You have got to pray.

Mr. POSEY. The upside, I guess, is that there is more public awareness now of the importance of space to the survival of our species and it is not at some unknown point in the far-distant future that we can imagine.

General BOLDEN. And sir, if I may, you said something that is so important. It would be very easy for this Congress and for the Administration to say—because we get the question all the time, why are we worried about exploring beyond low-Earth orbit, can't we just put that off for 5 or 10 years. The reason that I can't do anything in the next three weeks is because for decades we have put it off for the next 5 or 10 years. We don't have contractors who go away from doing their job and then 5 years from now we call and say okay, we want to build a rocket. They will tell me, with whom; we don't do that anymore. All those guys went over and they are now selling pizza, and I am not being facetious when I say that. And I apologize. You cause me to lose my temper sometimes when I—this is really important.

Mr. POSEY. Yes, it is.

General BOLDEN. And it has to be continuous. The President has a plan but that plan is incremental, and we can not like him, we can not agree with him, we can not do a lot of things. It is the best plan we have, and if we want to save the planet, because I think that is what we are talking about, then we have to get together, that side and that side, and decide how we are going to execute that plan as expeditiously as possible. That is all I can tell you.

Mr. POSEY. Thank you.

Chairman SMITH. Thank you, Mr. Posey. The gentleman from California, Mr. Takano.

Mr. TAKANO. Thank you, Mr. Chairman.

This use of the term ''civilization threatening'' or ''civilization destroying'' asteroids, remind me at what size would we say such an asteroid would be?

Dr. HOLDREN. A 1-kilometer asteroid would be carrying energy in the range of tens of millions of megatons. That is as much or more energy as was in the combined arsenals of the United States and the Soviet Union at the height of the Cold War. An asteroid of that size, a kilometer or bigger, could plausibly end civilization. Nobody has the detailed models, the ability to calculate and detail, to tell you exactly what the threshold is, but when you are talking about tens of millions of megatons of explosive energy, you are putting civilization at risk.

Mr. TAKANO. And I am hearing that we are relatively optimistic that we can develop systems at the right price points to be able to detect asteroids of this size with a sufficient amount of lead time to be able to do something about it.

Dr. HOLDREN. That is the size range where we have already detected something in the range of 93, 94 percent of the asteroids of that size range that could come close to the Earth, and in that size range, we can be reasonably assured, especially as we make these additional investments going forward, of being able to detect them with quite a lot of notice.

Mr. TAKANO. Let us scale it down to medium- to large-size city-destroying asteroids. What size would those be?

Dr. HOLDREN. A city-destroying asteroid could be in the range of 50-meter diameter carrying an energy in the range of 5 to 10 megatons.

Mr. TAKANO. What sort of systems would we need to be able to detect that? You talked about more assets in our orbit, telescopes of that kind including those that could get around the issue of the sun.

Dr. HOLDREN. We would want the infrared telescope in an orbit resembling that of Venus. It could be a Venus trailing orbit following the planet around, the planet Venus, which again is what the B612 Foundation is in fact working on. As Administrator Bolden mentioned, we actually had an experiment with an infrared telescope that was built for an orbiting telescope built for a different purpose. It is very good at finding asteroids.

Mr. TAKANO. We spoke a lot about the cooperative nature of what would need to happen, nations coming together, but would there be also rivalrous kinds of impulses which might divide us? In fact if we were to detect objects of this size, would nations also be concerned about that impacting the ability to detect missiles, for example?

Dr. HOLDREN. I think these are very different capabilities. As General Shelton mentioned, going into detail about our missile-detecting capabilities would require a different forum, but they are quite different in nature from the capabilities we would need to detect and track asteroids.

Mr. TAKANO. Well, the chairman raised a question that I thought was rather interesting, did none of our current missile-detecting capabilities, did they fail to be able to detect the most recent asteroid, and you may not be able to answer that question.

General SHELTON. I can. We did detect it, and as I said, it was at the time. It wasn't predicted. It was detection at the time.

Mr. TAKANO. So the missile detection capacities we have now I mean really are kind of—they are more in real time as opposed to time that we might be able to remediate the problem?

General SHELTON. Yes, sir, and focused on two things. The infrared signature coming out the back end of a missile, we see that, and as soon as it either breaks the ground, if there is weather overhead, as soon as it breaks the clouds, we will see that. We will be able to tell you what type of missile it is. We will be able to tell you where that missile is going. We will be able to tell you where it is going to impact. So very solid missile-warning capabilities. Those infrared sensors can be used for other things but they can't be used for predictive things out beyond Earth orbit.

Mr. TAKANO. Mr. Chairman, I am out of time. Thank you so much.

Chairman SMITH. Thank you, Mr. Takano. The gentleman from Arizona, Mr. Schweikert, is recognized.

Mr. SCHWEIKERT. Thank you, Mr. Chairman.

Just because I want to get my head around and try to really understand some of the base-level approach here, and Doctor, I was going to ask you first, and forgive me if I am equating a statement to you that was in someone else's opening statement. A dangerous interaction, Earth and an object, was the statement one-out-of-a thousand-year event?

Dr. HOLDREN. The one-in-a thousand-year event is the one of the magnitude that hit over the Tunguska, the asteroid impact over Siberia in 1908, and that was a 15-megaton class event. That is characteristic of one in a thousand years. The dimension of that asteroid was somewhere in the range of 50 meters.

Mr. SCHWEIKERT. Now, if I remember my old modeling classes, when you start getting into something with that far out in detail, you know, it is like the person that says it is a 500-year flood except we had three of them in the last 10 years, because you have such—your degree of confidence, your noise in that just becomes—it blows off the chart. So we always like to say one in a thousand but it is one in a thousand with, you know, a 20 percent lack of confidence. Does that sort of math also work for this?

Dr. HOLDREN. Well, I would say certainly there is a lack of confidence of that size or greater but the real catch is that a one-in-a-thousand-year event can occur at any time. The fact that on average one only expects these to happen once in a thousand years doesn't mean that one won't happen next year.

Mr. SCHWEIKERT. Often when we talk to certain non-statistical people, you try to explain that you can have the three 500-year floods in 10 years and then go 1,500 years without something.

Okay. In the discovery of objects out there, how much are you finding is coming from the amateur astronomy community? I mean, if I remember correctly, you were telling me that—was it the gentleman—was it an amateur in Spain that saw the last one?

Dr. HOLDREN. I am not sure it was an amateur.

General BOLDEN. I don't know that it was—we can find out whether it was an amateur astronomer. We just know it was an astronomer in Spain that made the discovery on 2012 DA14.

Mr. SCHWEIKERT. Is there—how formal or informal is that network out there of university amateurs, governmental astronomers, you know, scouring the skies, seeing things, reporting them? How does that mechanism work?

Dr. HOLDREN. It is actually quite organized, quite formal and quite fast. That community of folks stay in constant communication.

Let me take this opportunity to recommend a book, because it is not mine, a book by NASA's head of the near-Earth Object Identification program, Dr. Donald Yeomans. It just came out this year, 2013. It is called Near-Earth Objects: Finding Them Before They Find Us. Nice title. And he talks at great length about these networks, about the roles of amateurs, about the roles of professionals, who discovered what.

Mr. SCHWEIKERT. You are beating me into where I was actually trying to go. Is there a way to take that network and incentivize it? I have a great interest in sort of distributive information, distributive networks, so lots of smart people all over the world with this their hobbies, and is there a way—should we be incentivizing that?

Dr. HOLDREN. That is a great question, and we in OSTP are greatly in favor of crowdsourcing. We are greatly in favor of putting challenges out there, and in fact——

Mr. SCHWEIKERT. You and I are about to become really good friends.

Dr. HOLDREN. And these challenges we already know. We have used them across a domain of interesting problems, and I think there is no doubt we are going to have a challenge around asteroid detection.

Mr. SCHWEIKERT. And it is not answerable in 20-some seconds, but part of that is, okay, we see something. How far in advance with current technology do you have to see something to analyze, determine, you know, threat assessment and then react to it?

Dr. HOLDREN. The analysis and threat assessment is pretty fast because once you see it, you can train on it various other instruments—the radio telescopes, optical telescopes, and use the combination of information available from them once they know where to look in the sky to characterize its trajectory and determine whether or not it is a threat. The long-time scale, the long pole in the tent, is deploying the capability to deflect one that you discover is on a collision course, and that is the issue where currently we would have to say the time scale is in the range of years, and I think Administrator Bolden suggested that he would get back to the Committee on that, but I think his estimate, his initial estimate, is certainly reasonable. Even throwing a lot of resources at it, you would be talking 4 or 5 years to mount a deflection mission.

Mr. SCHWEIKERT. Mr. Chairman, thank you for your patience.

Chairman SMITH. Thank you, Mr. Schweikert. The gentlewoman from Connecticut, Ms. Esty, is recognized.

Ms. ESTY. Thank you very much, Mr. Chairman.

I too share some of the interest in this sort of crowdsourcing, and would just flag, since we have already had some hearings on big data, to perhaps follow up at a later time to think about what opportunities there are in other areas. We are also looking at the data side and how we might be able to collaborate on this worldwide problem, and I think that is very important.

For General Bolden, if you could talk a little bit about what NASA's procedure is for actually notifying our Federal agencies? You get notice of a NEO. What do you need to know? What triggers a notification warning and how does that actually work?

General BOLDEN. Congresswoman, there are several organizations we notify. We notify the State Department, first of all, because they notify our international partners that there is an incident, and this is not just for asteroids. This would be for a satellite that has fallen back to Earth or something, and we have had to exercise that several times over the last two years. The first person I would notify would be Dr. Holdren as the President's science advisor, and going back in response to Mr. Posey's question, there is no question in my mind who is in charge, and I go to Dr. Holdren because he pulls the team together, whether it is DOD or NASA and everyone else, but I understand the thrust of the question. So we would notify other Federal agencies, FEMA, the State Department, and then go from there. And it is scenario dependent. It depends on what the characterization of the asteroid or the NEO happens to be. Sometimes it is just a matter of saying hey, we now have something else that has been added to the inventory, it is not an Earth-threatening orbit, and we do that.

Ms. ESTY. Could you talk about whether there is an organized international warning network, or should there be? Is this something that is again scenario dependent or is there an actual formal network?

General BOLDEN. Dr. Holdren mentioned the recent meeting in conjunction with U.N. COPUOS that actually the chair was an American, a NASA scientist, and from that meeting came the initial decision that we would organize, and I can get you more information on what they propose, because like everything else, it is a proposal for an international collaborative effort to do this.

Dr. HOLDREN. If I could just add one thing to that. The Minor Planet Center, which I mentioned before, which is located at the Harvard-Smithsonian Astrophysical Observatory, is a formal international entity to which everybody automatically feeds discoveries of new near-Earth objects. So there is already a formal network which functions to assemble all the information that is available from all these different telescopes around the world, and even the amateur astronomers know where to go with their findings. They go straight to the Minor Planet Center, and the Minor Planet Center then goes to the NASA operation at JPL, which is responsible for working out the trajectory in coordination with these other groups. But the thing that is new, the international asteroid warning network, which emerged from this February 15th meeting in Vienna, will ramp up this whole effort and will add, I think, additional layers of capability as countries come together to say given these current scattered assets, what more do we need and how do we get it.

Ms. ESTY. It seems to me that is very important for several reasons. Everybody is under budget constraints so that we should be more effectively deploying world resources in this range but also confidence building, which I worry about from a security point of view, that if other countries see this as threatening because we might use these technologies in some other way, it is going to be vitally important that we are sharing in a way that in fact respects the assets other countries have and we all get the benefit for worldwide resources. So if you have specific proposals as the outcome of the Vienna conference goes forward, I hope you will come back to us to help us bring those forward to leadership about new opportunities but in fact will be lifesaving, you know, planet-saving potentially but that will allow—will require greater collaboration. Thank you very much.

Chairman SMITH. Thank you, Ms. Esty. The gentleman from Texas, Mr. Weber, is recognized.

Mr. WEBER. Thank you, Mr. Chairman.

Dr. Holdren, you said that the asteroid that hit Siberia was 15 megatons. What was the name of that event?

Dr. HOLDREN. That was Tunguska.

Mr. WEBER. Tunguska?

Dr. HOLDREN. T-u-n-g-u-s-k-a.

Mr. WEBER. Okay. And then you said, I think, you all agreed there was 13,000 objects——

Dr. HOLDREN. Thirteen to 20,000 140 meters and above, so the number would be somewhat larger for asteroids 100 meters and above.

Mr. WEBER. How close is the nearest one?

Dr. HOLDREN. Well, it is not a question of how close it is now. The question is, how close will its orbit take it to the Earth in the near future. Right now, as Administrator Bolden has said, none of these asteroids hat we have found is on a collision course with the Earth.

Mr. WEBER. Okay. You also—well, I think it was you, General Bolden, that said the Russian meteor was hidden by the sun and it is the reason we didn't detect it because it came straight out of the sun?

General BOLDEN. I wasn't, but that is correct. The folks in NASA when I asked the question of how did this happen, it came from out of the sun.

Mr. WEBER. But my question is, when something comes right out of the sun directly at us, at some point we are able to identify it, General Shelton, you said. How much time do we have? Is that 10 minutes, 2 hours? At what point does it become identifiable as it gets to the Earth's atmosphere?

General BOLDEN. Well, one thing, Congressman, I do have to re-emphasize, we talk about these three week scenarios, that is so unlikely, and even the occurrence in Russia, that was not a city-threatening—if you were in Russia, that was a significant event, but that is not of the size that is the city-threatening, the region-threatening, the other——

Mr. WEBER. But can you give me a time frame on how long we would have when one actually is in the——

General BOLDEN. It is my belief that we can identify in sufficient advance those that are the big threats, but we need to do better.

Mr. WEBER. Okay. We had the Hubble telescope up for a long time. Now we have replaced that——

General BOLDEN. It is still up.

Mr. WEBER. It is still up, and you and I had the discussion in my office, we have a better telescope up.

General BOLDEN. We are a little ways away. In 2018 we will launch the James Webb Space Telescope, but they are not in the asteroid NEO identifying—they are looking at totally different things.

Mr. WEBER. Given the scenario of low funding and time being of the essence, could we make that change to where we could add on to that telescope so we get it up in space?

General BOLDEN. No, sir. Very simply, no, sir.

Mr. WEBER. Can't do that?

General BOLDEN. No, sir. We would not want to do that, to be quite honest. We have a plan right now, Dr. Holdren and I both have mentioned collaboration with private industry, with private organizations like B612. I don't want anybody to think that B612 is going to save the planet but they are doing what we need to do in terms of providing a means to identify——

Mr. WEBER. That was my question about that particular telescope. The ISS, if I remember correctly, orbits the Earth every 91 minutes?

General BOLDEN. That is about right.

Mr. WEBER. How much of a role do they play in being able to identify and how much time do——

General BOLDEN. Right now we don't utilize it at all, but as I talked about when I was in with you, we are learning every single day that ISS, although we thought it was not a platform that you would want to do Earth science, it is turning out to be a great platform, and we are learning more and more about it. We have a solar experiment that is going up, and there may be the capability to put something there, but that is not going to be the answer.

Mr. WEBER. Six hours, six days, six weeks?

General BOLDEN. I would not even like to fool anybody that ISS and anything we can put on it is going to answer this question. The types of things that Dr. Holdren mentioned and I mentioned earlier are the way we need to go.

Mr. WEBER. All right. Two final questions and I have got to go. Who monitors this screen for all of these objects? Does it doing your iPhone when there is a threat coming? I mean, somebody has to got to be watching some instrument 24/7 to say oops, we picked one up. Who does that?

Dr. HOLDREN. That happens at the Minor Planet Center, where all the information from all of these sensing instruments around the world goes.

Mr. WEBER. And then final question. So you explode an asteroid, how do we know that we get total disintegration and we don't have, instead of one big object coming at us, 20 very lethal objects?

Dr. HOLDREN. You don't know that. That is one of the reasons that blowing one up close to the Earth is not a great option. Deflecting it farther from the Earth so that it doesn't hit us at all is a much better option.

Mr. WEBER. Thank you, Mr. Chairman. I yield back.

Chairman SMITH. Tank you, Mr. Weber. The gentleman from Texas, Mr. Veasey, is recognized.

Mr. VEASEY. Thank you. I forgot who it was earlier talked about an asteroid hitting an ocean and causing a tsunami. I guess depending on the size of the asteroid would be the correct answer to this question, but how far inland could a reasonably sized asteroid make water come in? Because that was really interesting to me.

Dr. HOLDREN. There is a very interesting discussion of exactly that question in Dr. Yeomans' book, and the answer is, we really don't know because the dynamics of tsunamis caused by asteroid impacts are, number one, very complicated and not adequately investigated, and it depends on many factors including the slope of the ocean bottom close to the continent that is going to be most affected and it depends on a lot of other characteristics of the asteroid impact. So I think there is no simple answer to that question that we can give at this time.

Mr. VEASEY. What about asteroids hitting other, you know, planet systems, or what sort of research do you have on that?

Dr. HOLDREN. Well, there are a lot of craters out there. There are craters on the moon from asteroid impacts that we can see very clearly.

Mr. VEASEY. Any recently that you—any recent craters on the moon?

Dr. HOLDREN. I would have to get back to you on that. I am not sure what the most recent impact on the moon is, but I think none very recent, but again, in geologic time, "recent" can be quite a stretch of time. But there is also lots of evidence of asteroids bashing into each other. If you look at the larger asteroids that are out there, they themselves are pitted with major craters that come from them bumping into each other.

Mr. VEASEY. Thank you.

Chairman SMITH. Thank you, Mr. Veasey. The gentleman from Utah, Mr. Stewart, is recognized.

Mr. STEWART. Thank you, gentlemen, for your time. I know you and your careers and I have a great deal of respect for you, so thank for that. General Bolden, good to see you, sir. We spent some time at your place talking the other day, and I know that you are a former Marine pilot. As you know, I am a former Air Force pilot.

My question is actually for General Shelton. As a senior Air Force officer with great wisdom and insight, is it your understanding, sir, as it is mine that Air Force pilots are the best pilots in the world?

General SHELTON. I am going to have to say yes on that, sir.

Mr. STEWART. Thank you, sir. I am surprised no one has asked that question yet. I am glad I was able to.

General SHELTON. Actually, sir——

General BOLDEN. That is fighter pilots of all services with the Air Force. I am an attack pilot.

Mr. STEWART. You are a bigger man than I am because I have never landed on a carrier.

Actually I have a couple of simple questions, then maybe a more detailed one. The first would be, you know, we spent a lot of time talking about detection avoidance, you know, and some of the uncertainties about that. I am curious about policy, public policy. If we were to determine that there was a threat and then even determined that it was actually potentially devastating, do we have a policy as to whether we would share that information with the public and how we would do that? And Dr. Holdren, I guess that is probably most appropriate for you.

Dr. HOLDREN. My expectation would be that we would notify, but the first thing that would happen if information came in indicating that an asteroid had been detected to be on a collision course with the Earth and it was big enough to do serious damage, it would be exactly what happened after the Fukushima earthquake and tsunami affected Japan. Namely, there would be a gathering in the Situation Room within minutes in which we would have the Chairman of the Joint Chiefs of Staff, we would have the Secretary of State, we would have the head of FEMA, we would have the Secretary of Homeland Security, we would have the head of NASA, we would have General Shelton, and there would be an intense discussion of the whole range of actions that the government would take in order to deal with the threat, whatever it was, and in that meeting, unquestionably there would be a discussion of who to notify, how fast, in what form.

Mr. STEWART. And I understand that. I am curious, and maybe—and I am not advocating one way or the other. I am just curious, have you determined the protocol for advising the public? Is that part of that matrix?

Dr. HOLDREN. I don't know whether FEMA, which would have that responsibility, has developed a formal protocol. We could get back to you on that.

Mr. STEWART. Okay. I wish you would. I would be curious to know that.

And the second thing, and we have all talked about it, maybe I am just not that bright, I am not sure I get it, but, you know, the saying, we don't know what we don't know, and you said that we have discovered 94 percent of the asteroids over 1 kilometer, for example, but if we don't know what is out there, how do we know that we have discovered 94 percent of them?

Dr. HOLDREN. That is actually a very good question, and it turns out that there are subtle statistical techniques that rely on sampling of subpopulations and what fraction of them you have see before in order to determine what fraction of the overall population you have actually seen. That is actually described again in wonderfully clear detail in Dr. Yeomans' book. It was the best explanation of that that I have seen.

Mr. STEWART. So you are interpolating there? You are drawing conclusions but you are fairly comfortable——

Dr. HOLDREN. You are drawing conclusions based on sampling.

Mr. STEWART. Right, but you are fairly comfortable with those figures?

Dr. HOLDREN. Yes.

Mr. STEWART. Okay. And then the last question in the minute or so I have left, you know, we talk about detection being, you know, the first line of defense, and our efforts, and you mentioned some of the others as well, but I mean, is the United States the lead on this? Clearly, we are, but are other nations contributing to this detection effort in a meaningful way or is it almost entirely our efforts that are meaningful here?

Dr. HOLDREN. No, absolutely other nations are contributing in a meaningful way. There are important telescopes and data centers in Italy. That is a German-Italian collaboration. There is another one in Czechoslovakia. There are some—the LSST will be in Chile. There are some in Australia. And again, this domain is actually remarkable for the degree of international cooperation and interconnection compared to many others where we are not nearly as far along.

Mr. STEWART. As it should be, of course, because we all got a dog in this fight. So those other entities, are they funded by the EU and other—they are not with American funding at all? Those are entirely independently funded efforts?

Dr. HOLDREN. No, they are not entirely independently funded. For example, the Large Synoptic Survey Telescope is being very substantially funded by NSF even though it is going to be in Chile, but of course, it will be an NSF facility in a sense. The Arecibo Radio Telescope in Puerto Rico is funded by NSF.

Mr. STEWART. So even though these are located, geographically located around the world, they are primarily U.S. efforts?

Dr. HOLDREN. I would have to get back to you on the international distribution of the funding. Certainly there is substantial funding from the European Space Agency. There is substantial funding from Germany, from Italy, from Czechoslovakia, from France, but I could not give you a percentage.

Mr. STEWART. Again, if you would, I would appreciate that. Thank you. And Mr. Chairman, I yield back. Thank you.

Chairman SMITH. Thank you, Mr. Stewart. The gentleman from Florida, Mr. Grayson, is recognized.

Mr. GRAYSON. Thank you, Mr. Chairman.

Gentlemen, we could spend each year a million dollars on space threats, we could spend a billion dollars or we could spend a trillion dollars. I would like to hear from each one of you what we should spend. That is what we have to decide here. And specifically, I would like to hear either a number or formula, I think the Science Committee can deal with formulas, or some sort of list of the things that you think must be done without regard to what they cost. Let us start with you, Dr. Holdren.

Dr. HOLDREN. The National Academy of Sciences just a couple of years ago came out with a report in which they actually addressed this question, and they looked at what you could do for $500 million a year, what you could do for $100 million a year, what you could do for 50. I would say on the basis of that, if we are just looking primarily at detection and characterization, that I think we would want to be spending upwards of $100 million a year. If we are looking, as I think we must as well, at mitigation, then you would have to include the costs of carrying out the President's goal of visiting an asteroid by 2025. Various estimates have been put forward of the cost of doing that, but it almost certainly would be in the range of $2 billion or more spread over the period between now and 2025.

Mr. GRAYSON. Thank you. General Shelton?

General SHELTON. Yes, sir. In my case, we are talking about geosynchronous orbit into the surface of the planet, so that, just that part of space that we are responsible for, probably 200 or 300 million a year-ish is what we are talking about, developing better sensors that are more sensitive to the space debris population that is growing, sensors that allow us to better catalog the activity that is there and characterize it as threats continue to grow in space both adversarial threats as well as environmental threats. We need to be able to characterize that much better than we have the capability to do today. So I would say that 200 to 300 million range is what we are talking about.

Mr. GRAYSON. Good. Administrator Bolden?

General BOLDEN. Sir, the only thing I will add, because Dr. Holdren pretty much answered it, I want to reemphasize, because we have identified 95 percent of those objects that are a kilometer and above and we have seen none that are on a collision trajectory with Earth, this is not an issue that we should worry about in the near term. However, as I said, the President has laid out a plan, and I would say that is a very good start. We have a lot of work to do but the funding that is presently laid out in the President's budget is sufficient to get us there incrementally. We just have to move that plan forward. So you can't stop. That is my point.

Mr. GRAYSON. All right. Now, tell us what kind of costs we would be facing if we spent nothing. It can be a worst-case scenario or a not-so-bad-case scenario, but the likely costs we would face if we did nothing. Let us start with you, Dr. Holdren.

Dr. HOLDREN. This is a very tough question because there are different ways to present these things. If you take the expected value of the damage in terms of loss of human life integrated over a very long period of time, it comes out that the estimated loss of life from asteroid impact is only about 100 per year. That compares with a million per year for malaria, it compares with five million per year for tobacco. So it doesn't look like a very big threat. But of course, that is not really a meaningful way to present a risk of this character where you are talking about a low probability of a very big disaster, and in those sorts of situations, we tend to invest in insurance to reduce the likelihood of a disaster we would regard as intolerable. If you say how big is the disaster, if you are talking about a 10-kilometer asteroid of the sort that exterminated the dinosaurs, what is the value of all of civilization? It is a very big number but is it meaningful as a number which you then divide by the 65-million-year return time? I think we just can't get at it that way.

Mr. GRAYSON. General Shelton, the costs of a worst-case scenario?

General SHELTON. Well, again, from a DOD perspective, we would not be able to characterize the traffic on orbit, we would not be able to avoid collisions on orbit, we would not be able to detect adversary activity on orbit, and our dependence on space, by the way, not only for our way of life but also for military operations is very high so we would sacrifice that.

Mr. GRAYSON. Thank you, Mr. Chairman.

Chairman SMITH. Thank you, Mr. Grayson, and let me thank our witnesses today for their testimony. This has been a particularly interesting hearing. No doubt there will be some follow-up questions that will be addressed to you all, but thank you for being here and thank you for your expertise as well.

We stand adjourned.

[Whereupon, at 11:58 a.m., the Committee was adjourned.]

Appendix I

ANSWERS TO POST-HEARING QUESTIONS

ANSWERS TO POST-HEARING QUESTIONS

Responses by The Honorable John P. Holdren
HOUSE COMMITTEE ON SCIENCE, SPACE, AND TECHNOLOGY

"Threats from Space: A Review of U.S. Government Efforts to Track and Mitigate Asteroids and Meteors, Part 1"

Questions for the Record, Dr. John P. Holdren, Director, Office of Science and Technology Policy
Executive Office of the President

Questions submitted by Rep. Steven Palazzo, Chairman, Subcommittee on Space

1. Why have NASA and the Administration not acted on the five recommendations on planetary defense made to the administrator in 2010 by the NASA Advisory Council's ad hoc Task Force on Planetary Defense?

   The five recommendations made by the Task Force were to: Organize for Effective Action on Planetary Defense; Acquire Essential Search, Tracking, and Warning Capabilities; Investigate the Nature of the Impact Threat; Prepare the Response to Impact Threats; and Lead U.S. Planetary Defense Efforts in National and International Forums. NASA has acted on these recommendations in several ways.

   The purpose of NASA's Near Earth Object Observations (NEOO) Program is to coordinate NASA-sponsored efforts to detect, track, and characterize potentially hazardous asteroids and comets that could approach the Earth. In FY 2010, the NASA budget for the NEOO Program was $5.8 million. The final report of the NASA Advisory Council Ad-Hoc Task Force on Planetary Defense was provided to the NASA Administrator in October 2010. NASA's budget formulation process for the subsequent President's Budget resulted in an Administration request of $20.4 million for FY 2012. The FY 2014 budget for NASA includes $40.5 million for the NEOO Program for near-Earth asteroid detection, follow-up, and characterization.

   Within the increased investment in the NEOO Program, the 2014 Budget includes focused support for partnerships and leveraging, including international and commercial partnerships. The objectives of the enhanced NEOO Program are responsive to the Task Force's recommendations to Acquire Essential Search, Tracking, and Warning Capabilities; and Investigate the Nature of the Impact Threat.

   In addition, NASA has taken specific action to strengthen the leadership of U.S. planetary defense efforts in national and international forums. The NASA NEOO Program has provided essential leadership to the United Nations Committee on Peaceful Uses of Outer Space's Science and Technology Subcommittee's action team on the NEO threat. Follow-on work will continue in international fora to develop a plan for an enhanced international asteroid warning network, impact disaster planning, space-mitigation mission planning should there be a credible threat, and advice on planning and response.

   To "Prepare the Response to Impact Threats," on April 3, 2013, the NEOO Program and the Department of Homeland Security's Federal Emergency Management Agency (FEMA) held a one-day simulation of impact disaster response with NEO observation, detection, and mitigation community experts and FEMA emergency-response personnel. Also, the Planetary Defense Conference, held this April in Flagstaff, AZ, conducted a half-day international impact-emergency-response exercise with participants from multiple

countries.

2. How often do we currently observe large meteors entering the atmosphere safely over the ocean?

   Every day, a continual influx of meteors strikes Earth's atmosphere. Most of them are dust-sized particles, but they add up; it's been estimated that on a typical day, these particles total from 50 to 150 tons of matter. Asteroids of the order of a few meters in size strike the atmosphere roughly annually. About 70 percent of the Earth's surface is covered by water, and asteroids do not originate from any preferred direction in the sky, so we expect that the majority of these annual impacts by meter-sized asteroids take place over the oceans. Larger asteroid impacts are even less frequent; the probability of an asteroid as large as 140 meters in diameter striking the Earth is estimated at 1 in 30,000 per year.

3. What capabilities does the U.S. government already possess to detect and track asteroids? What level of fidelity is needed compared to the level of fidelity we have?

   NASA sponsors a number of activities relating to the search for NEOs under its Near Earth Object Observation (NEOO) program, including work at the international Minor Planet Center (MPC), located at the Harvard-Smithsonian Center for Astrophysics, which collects and correlates NEO orbit data; research at two radio-telescope facilities that help provide precision tracking and characterization of NEOs; surveys conducted by ground-based optical telescopes; and activities at the NASA NEO Program Office at the Jet Propulsion Laboratory (JPL), which coordinates assessments of NEO orbits and impact probabilities. There are also cooperative projects involving NASA, the National Science Foundation (NSF, which has a key role within the United States for ground-based astronomical assets), and the U.S. Air Force (USAF) Panoramic Survey Telescope and Rapid Response System (PanSTARRS) program, as well as non-government academic and space research organizations. Additionally, NEO detection is a major science driver for the proposed Large Synoptic Survey Telescope (LSST). NASA is also working with the Canadian Space Agency (CSA) on processing of data that will be collected from the CSA Near-Earth Object Surveillance Satellite (NEOSSat) launched in February 2013.

   These assets constitute an effective program for discovering larger NEOs, but we need to improve our capabilities for the identification and characterization of smaller NEOs. Specifically, with our current or near-future capabilities, both on the ground and in space, it is unlikely that objects smaller than 100 meters in diameter on collision courses with the Earth will be detected with more than weeks of advance warning – a matter of some concern since the larger objects in this range could be city-destroyers. Small NEOs are difficult to detect in visible light from ground-based telescopes because the small, dark objects reflect only a small amount of visible sunlight. In contrast, telescopes sensitive to infrared light detect an object's heat, rather than reflected sunlight; even small, dark asteroids could be detected by a telescope sensitive to infrared light, making these capabilities particularly relevant for future NEO surveys.

Questions for the Record
Ranking Member Eddie Bernice Johnson

"Threats from Space: A Review of U.S. Government Efforts to Track and Mitigate
Asteroids and Meteors, Part 1"

March 19, 2013

1. The B612 organization's space-based NEO detection telescope initiative (Sentinel) was mentioned during the hearing. What efficiencies, if any, in cost and in the detection of incoming NEOs, would Sentinel provide as compared to alternative approaches?

    The Sentinel mission concept is to employ a space-based infrared telescope on a satellite in a Venus-trailing orbit in order to detect and track asteroids from a vantage point that is very conducive to identifying potentially hazardous objects (PHOs) approaching Earth. Current analysis of the Sentinel capabilities indicate that it may be 100 times better at identifying near-Earth asteroids than all ground-based telescopes combined (not including the Large Synoptic Survey Telescope {LSST}). If the LSST is considered, this single spacecraft mission would be about 10 times better than all ground-based capabilities combined. Other space-based capabilities are being explored, including reactivation of NEO WISE (a previous secondary tasking of the WISE spacecraft), Canada's low-Earth orbit (LEO) NEOSSat, consideration of hosted payloads on spacecraft in geosynchronous (GEO) orbit, and the Jet Propulsion Laboratory's (JPL) study work on the NEOCAM system. Space-based assets have the advantage of eliminating atmospheric aberrations compared to their ground-based telescope counterparts.

    a. What, if any, role does the Administration anticipate NASA having in this initiative?

        NASA currently has a nonreimbursable Space Act Agreement with the B612 Foundation for the Sentinel mission, signed in June 2012. A NASA Technical Consulting Team was established to support the B612 Project Concept and Integration Review (PCIR). NASA will also provide B612 access to its Deep Space Network for telecommunications with the Sentinel spacecraft for commanding and data downlink.

    b. Does NASA plan to obtain data from this organization's telescope? What would be the cost to NASA of obtaining such data?

        Per the nonreimbursable Space Act Agreement between NASA and the B612 Foundation, NASA will receive all the data on a no-exchange-of-funds basis in return for technical and operational assistance during development and operations of the Sentinel spacecraft. NASA estimates the cost of the support provided to B612 over the life cycle of Sentinel to be about $30 million.

    c. What are the Administration's plans should the B612 organization not obtain the private funding needed to develop its space telescope?

        The Administration and NASA are pursuing a broad, diverse, and balanced set of approaches to detecting, characterizing, and tracking asteroids, of which the B612 organization is one important contributor. Lack of private funding to support key

development milestones would have a negative impact on the ability of both the B612 Foundation and NASA to leverage this important element of the portfolio in a timely and efficient manner. If B612 doesn't obtain the necessary funding, other solutions will be explored, including greater public-private cooperation, interagency and/or international engagement, alternative mission concepts (internal or external to NASA), and de-scoping or delays to planned capability until the private funding is in place.

2. According to NASA's Near Earth Object Program webpage, *"While striking the early Earth billions of years ago, comets are thought to have created major changes to Earth's early oceans, atmosphere, and climate, and may have delivered the first carbon-based molecules to our planet, triggering the process of the origins of life."* What changes to the Earth system would be expected if an asteroid or comet were to impact Earth in the next century?

The effect of an asteroid or comet impact is largely dependent on the size of the asteroid and where it strikes the Earth. For example, many small meteors strike the Earth's atmosphere and disintegrate entirely high in the atmosphere, never impacting the surface of the Earth. The 1908 explosion over Tunguska in Siberia, which leveled trees over an area of more than 2,200 square kilometers (850 square miles), is thought to have been caused by a stony asteroid between 45 and 60 meters in diameter, imparting between 10 and 20 megatons of TNT equivalent. An energy release of this size could cause hundreds of thousands of casualties and massive destruction if it occurred over an urban area. A similarly sized or even larger asteroid that made it to the surface intact could cause significant damage even if it hit the ocean, by virtue of the resulting tsunami. Depending on its composition and velocity, an asteroid of 140 meters in diameter could have an impact energy in the range of 50 to 500 megatons of TNT equivalent and would be capable of causing destruction over a large region. Impacts from massive objects that are several kilometers in size would have the most devastating worldwide effects. These impacts would include massive tsunamis if the object strikes the ocean and a massive cloud of dust and ash if it strikes land. The dust and ash would rapidly spread throughout the Earth's atmosphere, altering the temperature of the planet and preventing sunlight from reaching the surface, possibly for many years to come.

3. The Committee has been in the forefront of U.S. policy on near-Earth object detection and monitoring for the past two decades. What priorities regarding U.S. government efforts in detection, tracking, and mitigation of potentially hazardous near-Earth objects as well as leading and effecting emergency response and preparedness should this Committee be considering as it discussed NASA reauthorization this year?

Detecting, tracking, and mitigating potentially hazardous near-Earth objects have been priorities during this Administration. The President's 2014 Budget Request for NASA's Near Earth Object Observation Program proposed more than a ten-fold increase in funding (to $40.5 million from $4 million) from the 2009 funding level for NEO detection activities. Further, the President's National Space Policy specifically directs NASA to "pursue capabilities, in cooperation with other departments, agencies, and commercial partners, to detect, track, catalog, and characterize near Earth objects to reduce the risk of harm to humans from an unexpected impact on our planet and to identify potentially resource-rich planetary objects." This guidance also reinforces NASA's roles and responsibilities with regard to NEOs, as well as those of other Federal departments and agencies including the

Department of Defense, the Department of State, and the Department of Homeland Security's Federal Emergency Management Agency (FEMA).

OSTP has been working closely with several departments and agencies to draft plans and procedures, including potential mitigation strategies, that could be used in the unlikely event of an impending NEO impact threat. Under these plans, it is NASA's responsibility to provide initial notice of such a threat. Following such notification, communications resources and mechanisms already in place within FEMA would be used to communicate information domestically. The Department of State's diplomatic mechanisms would come into play for international communications as needed. DOD and NASA have already shown leadership by taking the initiative to run multi-agency disaster and deflection exercises, and by collaborating in the development of an international disaster and deflection response scenario for the recent Planetary Defense Conference hosted by the International Academy of Astronautics in Flagstaff, Arizona.

4. What are the Administration's plans for addressing policy and legal issues regarding NEOs-e.g., when and how to warn the public and whether to use nuclear explosives to detect an asteroid- be handled on national and international levels? What steps has OSTP taken to date to address such issues and what further actions will be pursued?

The Executive Office of the President, in conjunction with Federal departments and agencies, is reviewing and assessing the legal and policy issues related to Near Earth Objects, in both public and private fora. Potential asteroid-impact mitigation strategies span a wide spectrum of options, with several study and analysis efforts ongoing both within NASA and across the broader community. OSTP provided a letter report to Congress in October 2010 outlining the broad issues related to planetary defense from asteroids. Integrated efforts continue, including the International Planetary Defense Conference on 15-16 April, 2013 in Flagstaff, AZ, which included several interagency, industry, academia, and international components, and the bilateral NASA-FEMA Table Top Exercise on Planetary Defense, Warning, and Communication held in Washington on April 3, 2013.

5. The Ad-Hoc Task Force on Planetary Defense of the NASA Advisory Council recommended in 2010 that the cost of NASA Planetary Defense activities be explicitly budgeted by the Administration and funded by the Congress as a separate agency budget line, not diverted from existing NASA science, exploration, or other mission budgets. Does the Administration plan to do so? If not, why not? What types of analyses would need to be completed before a credible cost estimate for the activities recommended by the Task Force could be provided to Congress?

NASA's NEOO Program is already budgeted as a distinct line item within the Science Mission Directorate's Planetary Science budget. Pursuant to Section 321 of the NASA Authorization Act of 2005 (Public Law No. 109-155), the NASA Administrator transmitted a report to Congress in March 2007, entitled "Near-Earth Object Survey and Deflection Analysis of Alternatives." This report included very preliminary cost estimates for select architectures. The Ad Hoc Task Force incorporated that analysis into its review, but since over 5 years have now passed, a similar cost analysis would be necessary to produce a credible cost estimate.

Questions for the Record
Representative Donna F. Edwards

"Threats from Space: A Review of U.S. Government Efforts to Track and Mitigate Asteroids and Meteors, Part 1"

March 19, 2013

1. How important is the proposed human mission to a near-Earth asteroid to the Administration's overall planning for mitigation of a potential NEO threat to the United States? Given the timeframe in which the human NEO effort is proposed (2020s) and the challenges that remain for such a mission, what additional or alternative approaches will the Administration pursue in order to inform mitigation decisions and plans that could maximize the preparedness of the United States and its international partners at the earliest possible date?

As part of the agency's overall asteroid strategy, NASA is planning a first-ever mission to identify, capture, and redirect an asteroid into a stable orbit around the Earth about 40,000 miles outside that of the Moon. The overall mission is composed of three separate and independently compelling elements: the detection and characterization of candidate near-Earth asteroids; the robotic rendezvous, capture, and redirection of a target asteroid to the Earth-Moon system; and a crewed mission to explore and sample the captured asteroid using the Space Launch System (SLS) and the Orion crew capsule. Accomplishing this mission would represent an unprecedented technological achievement – raising the bar for human exploration and discovery, while demonstrating capabilities needed to protect our home planet and bringing us closer to a human mission to Mars in the 2030s.

The proposed human mission to a near-Earth asteroid is not being undertaken <u>primarily</u> for planetary-defense purposes, but rather to advance exploration capabilities and technologies while bringing side benefits in the planetary-defense domain. As part of the proposed mission, NASA is proposing to double funding for NEO detection, which will assist both in detecting a suitable target asteroid for the mission and in enhancing detection of potentially hazardous NEOs. NASA's enhanced NEO Observation Program will continue in parallel with the human asteroid mission and will take advantage of synergies that could benefit both efforts. The planetary defense portion of the asteroid mission will also utilize innovative methods (*e.g.* public-private partnerships, citizen science and crowdsourcing, prizes and challenges, etc.) to engage national and international partners to explore detection, tracking, characterization, and mitigation solutions for potentially hazardous NEOs.

2. Since your October 2010 response to Congress, has anyone assessed the effectiveness of communication and coordination among U.S. government agencies involved in near-Earth object surveying, detection, and characterization? What, if any, improvements are needed and how does the Administration plan to address those needs?

Several Federal departments and agencies have significant roles in the pursuit of these goals, and they cooperate in important ways. NASA sponsors various activities relating to the search for NEOs, including the collection and correlation of NEO orbit data, precision tracking and characterization of NEOs, and assessments of NEO orbits and impact probabilities in conjunction with other U.S. government agencies including the Department of Defense and U.S. Air Force, and the National Science Foundation (NSF), each of which

plays a key role in funding ground-based astronomical assets that are used to detect and track NEOs. In addition, the Minor Planet Center, located at the Harvard-Smithsonian Astrophysical Observatory, is a formally constituted international entity to which both professional and amateur astronomers feed discoveries of new near-Earth objects for follow-up and tracking.

In reports issued in 2010, both the National Research Council and the NASA Advisory Council examined the NEO issue. While the focus of their recommendations varies, both groups agreed on the importance of Congressionally-directed efforts to detect, track, and characterize potentially dangerous asteroids. Both also point out that additional investments in technology and hardware are required to achieve the NEO-search goals in a timely manner. The President's FY 2014 Budget Request includes $40.5 million for NASA's NEOO Program to continue to improve NEO detection capabilities, a ten-fold increase over the 2009 budget of $4 million just five years ago.

3. How well understood are the potential approaches to deflecting asteroids? What is the confidence level in the technologies that would be required? What information is needed to assess the various approaches, and how will decisions be made on which mitigation strategy to take?

The most effective approach to mitigation of a potential asteroid impact threat is highly dependent on the scenario. Near-term impact of an asteroid tens of meters in diameter requires a significantly different approach than the threat of a larger object that might impact decades in the future. The orbit parameters of the potential impactor are also a significant factor in determining an effective mitigation strategy. Therefore, a "toolkit" of mitigation approaches needs to be developed at the conceptual level to address the range of potential impact threats. While considerable thinking has gone into a variety of approaches, more extensive analysis will be required before any of them can be considered well understood. As a next step, in FY 2014, NASA's Office of the Chief Technologist plans to develop a "roadmap" of mitigation technologies.

4. To what extent do international space agencies or international facilities contribute to NASA's NEO survey and/or a worldwide effort of surveying, tracking, and characterizing potentially hazardous near-Earth objects? Since your October 2010 response to Congress, has the effectiveness of communication and data-sharing on near-Earth object tracking among nations been assessed? If not, are there plans to do so?

NASA has taken specific action to strengthen the leadership of U.S. planetary defense efforts in national and international forums. The NASA NEOO Program has provided essential leadership to the United Nations Committee on Peaceful Uses of Outer Space's Science and Technology Subcommittee's action team on the NEO threat. Follow-on work in international fora will continue to develop a plan for an enhanced international asteroid-warning network, impact disaster planning, space-mitigation mission planning should there be a credible threat, and advice on planning and response. NASA is also working with the Canadian Space Agency (CSA) on processing of data that will be collected from the CSA Near-Earth Object Surveillance Satellite (NEOSSat) launched in February 2012. There has not been a formal assessment of international efforts in near-Earth object tracking since the October 2010 report to Congress.

5. What is the approximate cost for the U.S. government to have a serious near earth asteroids program and which agency should bear the brunt of that expense?

The purpose of NASA's Near Earth Object Observations (NEOO) Program is to coordinate NASA-sponsored efforts to detect, track and characterize potentially hazardous asteroids and comets that could approach the Earth. In FY 2010, the NASA budget for the NEOO Program was $5.8 million. The FY 2014 budget for NASA includes $40.5 million for the NEO Program for near- Earth asteroid detection, follow-up and characterization. This increased funding is in support of human exploration as well as to protect our planet.

   a. NASA is the science agency, yet the Department of Defense is responsible for national defense – would this be a shared responsibility?

   The President's National Space Policy specifically directs NASA to "pursue capabilities, in cooperation with other departments, agencies, and commercial partners, to detect, track, catalog, and characterize near Earth objects to reduce the risk of harm to humans from an unexpected impact on our planet and to identify potentially resource-rich planetary objects." This guidance also reinforces NASA's roles and responsibilities with regard to NEOs, as well as those of other Federal departments and agencies including the Department of Defense, the Department of State, and the Department of Homeland Security's Federal Emergency Management Agency (FEMA).

   OSTP has been working closely with several departments and agencies to draft plans and procedures, including potential mitigation strategies, that could be used in the unlikely event of an impending NEO impact threat. Under these plans, it is NASA's responsibility to provide initial notice of such a threat. Following such notification, communications resources and mechanisms already in place within FEMA would be used to communicate information domestically. The Department of State's diplomatic mechanisms would come into play for international communications as needed. DOD and NASA have already shown leadership by taking the initiative to run multi-agency disaster and deflection exercises, and by collaborating in the development of an international disaster and deflection response scenario for the recent Planetary Defense Conference hosted by the International Academy of Astronautics in Flagstaff, Arizona. DOD could play an important role in the event of an impending NEO impact threat, depending on the approach deemed most appropriate for dealing with it

   b. When could we reasonably expect such a program to come on line – of course, given the appropriate resources were so allocated?

   As described in the testimony and in the earlier responses, elements of a near-earth asteroid mitigation program are already underway but at this time we are not in a position to provide a timeline nor a determination of what resources would be needed for a mitigation program that would actively defend from all potential asteroid threats.

6. What is the likelihood of such a devastating impact from a near earth asteroid? Are the resources necessary for a credible program worth the cost given the low probability of such an occurrence in our lifetimes?

   As indicated in my written testimony, the likelihood of a devastating impact from a near-

Earth asteroid is very low when expressed as an annual probability. Nonetheless, improbable events do occur, and if the potential consequences for society are large – as is the case for impacts of asteroids in the 100-meter or larger size range – it is prudent to invest in strategies for early warning and mitigation.

Responses by Gen. William L. Shelton
CHARRTS No.: HSSTC-01-001
Hearing Date: March 19, 2013
Committee: HSSTC
Member: Congressman Palazzo
Witness: Gen Shelton
Question: #1

Question: Many of our space assets are also extremely vulnerable to NEOs. Given our reliance on these assets, what plans are in place to mitigate potential damage to our satellites and International Space Station? a. Do our current capabilities provide adequate tracking and warning of potential harmful impact to our space assets? b. What protocol has been established for giving advanced warning to U.S. government assets and are these also provided to the international community or commercial operators?

Answer: Joint Functional Component Command for Space (JFCC SPACE) sensors, command and control, and analysis functions are designed to track man-made objects in Earth orbit. They track objects from low Earth orbit (LEO) through medium Earth orbit (MEO) to geosynchronous Earth orbit (GEO). They do not detect or track non-Earth orbiting objects. When alerted by NASA of an NEO that is expected to come close enough to the earth to threaten a satellite, including the International Space Station, JFCC SPACE will use NASA generated positional data to screen all active satellites for possible collisions. The Joint Space Operations Center (JSpOC) can then notify the owner/operators of those satellites of the threat. During the recent asteroid close approach, NASA detected and tracked that object. NASA then converted their data on the object to a format JFCC SPACE/JSpOC could use in its systems. This allowed JSpOC to perform conjunction assessment to determine if any assets were at risk (they were not).

a. Do our current capabilities provide adequate tracking and warning of potential harmful impact to our space assets?

Answer: The JSpOC is 100% reliant on outside agencies such as NASA to identify non-manmade objects that will come close enough to the earth to threaten satellites. JFCC SPACE has no ability to track or warn of these potential threats. There is, however, a working relationship between the JSpOC and NASA for screening these threats when identified. Only NASA will be able to determine if their tracking and warning system is adequate. Again, JFCC SPACE detects and tracks man-made objects in Earth orbit, and requires NASA notification and NASA-collected data to have any opportunity to provide an assessment of risk to space assets.

b. What protocol has been established for giving advanced warning to U.S. government assets and are these also provided to the international community or commercial operators?

Answer: JFCC SPACE has 38 Space Situational Awareness agreements with owner/operators of satellites from many different countries. They coordinate with the State Department to release information to some foreign entities and share much of their information through the Spacetrack.org website. There are over 50,000 users of that website. This protocol is designed to report when there are possible conjunctions between Earth-orbiting man-made objects. JFCC SPACE could use it to report possible conjunctions between NASA-reported natural objects, but only if NASA provided the notification as well as data on the object that JFCC SPACE could use.

CHARRTS No.: HSSTC-01-002
Hearing Date: March 19, 2013
Committee: HSSTC
Member: Congressman Palazzo
Witness: Gen Shelton
Question: #2

Question: Can you describe to us the details of the recently signed Memorandum of Agreement between Air Force Space Command and NASA's Science Mission Directorate from January 18 of this year?

Answer: The classified Memorandum of Agreement describes the process for the release of data on bolides (meteoric fireballs) to NASA, and it describes the specifics of what data can be provided and the format used to provide this data.

CHARRTS No.: HSSTC-01-003
Hearing Date: March 19, 2013
Committee: HSSTC
Member: Congressman Palazzo
Witness: Gen Shelton
Question: #3

Question: How often do we currently observe large meteors entering the atmosphere safely over the ocean?

Answer: Air Force Space Command only tracks man-made objects in Earth orbit. This question is in the purview of NASA.

CHARRTS No.: HSSTC-01-004
Hearing Date: March 19, 2013
Committee: HSSTC
Member: Congressman Palazzo
Witness: Gen Shelton
Question: #4

Question: What capabilities does the U.S. government already possess to detect and track asteroids? What level of fidelity is needed compared to the level of fidelity we currently have?

Answer: Air Force Space Command only tracks man-made objects in Earth orbit. This question is in the purview of NASA.

CHARRTS No.: HSSTC-01-005
Hearing Date: March 19, 2013
Committee: HSSTC
Member: Congressman Palazzo
Witness: Gen Shelton
Question: #5

Question: What unique characteristics can the U.S. Space Command offer to the overall asteroid detection and disaster mitigation program? How are these activities related to general space debris mitigation activities?

Answer: Air Force Space Command equips United States Strategic Command (USSTRATCOM) to detect and track man-made Earth orbiting objects using the Space Surveillance Network (SSN). If queued, and the asteroid or meteor is close enough, our sensors can detect and track. However, there is very limited capability to create orbits from these hyperbolic tracks. For general space debris mitigation, if NASA provided well defined trajectory and daily ephemeris data for an asteroid, the Joint Functional Component Command for Space's (JFCC SPACE) Joint Space Operations Center (JSpOC) could provide satellite conjunction assessment to identify any close approaches with man-made satellites.

CHARRTS No.: HSSTC-01-006
Hearing Date: March 19, 2013
Committee: HSSTC
Member: Congresswoman Edwards
Witness: Gen Shelton
Question: #6

Question: The National Research Council (NRC) reported in 2010 that data from DoD operated sensors in Earth orbit are capable of detecting the high-altitude explosion of small NEOs and that DoD had shared this information with the NEO science community in the past. The NRC report included a recommendation that data from NEO airburst events observed by the U.S. Department of Defense satellites be made available to the scientific community. What are the pros and cons from your perspective of sharing data from NEO airburst events observed by DoD satellites and ground sensors with the scientific community?

Answer: We are unaware of any specific benefit to the DoD for reporting high-altitude explosions of small NEOs. However, we understand that the scientific community finds this type of data helpful in identifying NEO objects and locating meteoroids that make it to Earth's surface. Air Force Space Command does have an agreement to share some data on NEOs with NASA. However, details are classified in order to protect military capabilities.

CHARRTS No.: HSSTC-01-007
Hearing Date: March 19, 2013
Committee: HSSTC
Member: Congresswoman Edwards
Witness: Gen Shelton
Question: #7

Question: You state in your prepared statement that the Joint Space Operations Center (JSpOC) has recently assisted NASA in analyzing the orbit data from asteroid 2012 DA 14 which recently flew by Earth. You further describe current JSpOC data processing capacity limitations and ongoing efforts to field the next generation system. With the increased automation being made available, will JSpOC be able to provide NASA with a greater level of support in the near future, and if so, what kind of support?

Answer: The first increment of the next generation system (JSpOC Mission System or JMS) became operational this year. The major capabilities now available to NASA include improved access to the space operational picture and basic space services over the SECRET Internet Protocol Router Network (SIPRNET). The next major increment delivered will provide NASA access to significant improvements in space situational awareness in several areas. First, the JSpOC will be able to provide NASA with near-real-time, highly accurate positional information on all satellites and debris tracked by our Space Surveillance Network (SSN). Second, the JSpOC will provide NASA with conjunction assessment calculations on all possible conjunctions between manned missions, NASA satellites, and all other tracked objects in Earth orbit. Note that the primary limitation of this reporting is based on the capability of our SSN to track objects and not on the capacity of JMS. Future increments will allow JMS to use NASA-supplied asteroid track data to calculate potential conjunctions between the asteroid and satellites in Earth orbit.

CHARRTS No.: HSSTC-01-008
Hearing Date: March 19, 2013
Committee: HSSTC
Member: Congresswoman Johnson
Witness: Gen Shelton
Question: #8

Question: To what extent has DoD been involved in steps to address policy and legal issues relating to NEOs, e.g., when and how to warn the public and whether to use nuclear explosives to deflect an asteroid?

Answer: Policy and legal issues relating to NEOs generally have not been part of the DoD dialog. DoD has been much more concerned with activity in earth orbit.

CHARRTS No.: HSSTC-01-009
Hearing Date: March 19, 2013
Committee: HSSTC
Member: Congressman Neugebauer
Witness: Gen Shelton
Question: #9

Question: What is the likelihood of such a devastating impact from a near earth asteroid? Are the resources necessary for a credible program worth the cost given the low probability of such an occurrence in our lifetimes?

Answer: Air Force Space Command only tracks man-made objects in Earth orbit, and therefore has not been part of the discussion on likelihood of impacts by asteroids, nor has the Command been part of the discussion on programs for detection of asteroids.

Responses by The Honorable Charles F. Bolden, Jr.
HOUSE COMMITTEE ON SCIENCE, SPACE, AND TECHNOLOGY

"Threats from Space: A Review of U.S. Government Efforts to Track and Mitigate Asteroids and Meteors, Part 1"

Questions for the Record, General Charles F. Bolden, Jr., Administrator
National Aeronautics and Space Administration (NASA)

Questions submitted by Rep. Steven Palazzo, Chairman, Subcommittee on Space

1. Many of our space assets are also extremely vulnerable to NEOs. Given our reliance on these assets, what plans are in place to mitigate potential damage to our satellites and the International Space Station?

    a. Do our current capabilities provide adequate tracking and warning of potential harmful impact to our space assets?
    b. What protocol has been established for giving advanced warning to U.S. government assets and are these also provided to the international community or commercial operators?
    c. How often do you have to alter the path of the ISS to avoid a possible debris strike? Is a lot of propellant used in doing so?

    ANSWER: It is highly unlikely that the International Space Station (ISS) or other assets in space would be struck by a sizeable NEO; however, space-based assets are hit by micrometeoroids and very small space debris frequently. The Joint Space Operations Center (JSpOC) and the NASA Trajectory Operations Center (TOPO) teams continuously monitor for potential collisions with debris of any significant size, and provide adequate notification so that avoidance maneuvers can be planned and executed.

    A total of 15 Debris Avoidance Maneuvers (DAMs) have been performed by ISS and the Space Shuttle from 1998 to present. Of those, the ISS has performed 11 and the Space Shuttle performed 4 while attached to ISS. A typical DAM would use about 75 kilograms of propellant.

2. Does NASA have the ability to track objects that could potentially be harmful to astronauts engaged in deep space exploration?

    ANSWER: The risk of impact from a NEO of the type being tracked by NASA to a deep-space mission is considered minimal. Though the risk is also very limited, spacecraft on deep-space missions would more likely encounter micrometeoroids, and these would not be tracked.

3. Please provide details of the recently signed Memorandum of Agreement between Air Force Space Command and NASA's Science Mission Directorate from January 18, 2013.

ANSWER: The recently signed MOA will support the NASA NEO program's ability to share information with the scientific community about fireball and bolide reports. Fireballs and bolides are astronomical terms for exceptionally bright meteors that are spectacular enough to be seen over a very wide area. The NEO program provides a chronological data summary of the brightest fireball and bolide events provided by U.S. Government sensors; this data summary can be accessed at http://neo.jpl.nasa.gov/fireballs/.

4. How often do we observe large meteors entering the atmosphere safely over the ocean?

   ANSWER: Every day, a continual influx of meteors strikes Earth's atmosphere. Most of them are dust-sized particles, but it has been estimated that on a typical day, these particles total from 50 to 150 tons of matter. Asteroids of the order of a few meters in size strike the atmosphere roughly annually. About 70 percent of the Earth's surface is covered by water and asteroids do not originate from any preferred direction in the sky, so we expect that the majority of these annual impacts by meter-sized asteroids take place over the oceans. Larger asteroid impacts are even less frequent; an asteroid as large as 140 meters in diameter striking the Earth is estimated to average about 1 in 30,000 years.

5. What capabilities does the U.S. government already possess to detect and track asteroids? What level of fidelity is needed compared to the level of fidelity we currently have?

   ANSWER: NASA sponsors a number of activities relating to the search for NEOs under its Near Earth Object Observation (NEOO) program, including: work at the international Minor Planet Center (MPC), located at the Harvard-Smithsonian Center for Astrophysics, which collects and correlates NEO orbit data provided by observatories around the world; surveys conducted by several search teams operating ground-based optical telescopes; activities at the NASA NEO Program Office at the Jet Propulsion Laboratory (JPL), which coordinates assessments of NEO orbits and computes impact probabilities; and, research at two radio-telescope facilities that provide precision tracking and characterization of NEOs. There are also cooperative projects involving NASA, the National Science Foundation (NSF), which has a key role within the United States for ground-based astronomical assets, and the U.S. Air Force (USAF) Panoramic Survey Telescope and Rapid Response System (PanSTARRS) program, as well as non-government academic and space research organizations. Additionally, NEO detection is a major science driver for the proposed Large Synoptic Survey Telescope. NASA is also working with the Canadian Space Agency (CSA) on processing of data that will be collected from their recently launched Near-Earth Object Surveillance Satellite (NEOSSat).

   These assets constitute an effective program for discovering larger NEOs, and we are working to improve our capabilities for the identification and characterization of smaller, few hundred meter sized NEOs. Small NEOs are difficult to detect in visible light from ground-based telescopes because the small, dark objects reflect only a small amount of visible sunlight. In contrast, telescopes sensitive to infrared light detect an object's radiated heat, rather than reflected sunlight; even small, dark asteroids could be detected by a telescope sensitive to infrared light, making these capabilities particularly relevant for future NEO surveys. However, these sensors must operate outside the Earth's atmosphere to be effective.

6. Please provide a status update on the activities of NASA's Near Earth Objects Office.

    ANSWER: The purpose of NASA's Near Earth Object Observations (NEOO) Program is to coordinate NASA-sponsored efforts to detect, track, and characterize potentially hazardous asteroids and comets that could approach the Earth. As noted in the response to Question 5, the NEOO Program continues to support a number of activities relating to the search for NEOs with our partners. In particular, the Minor Planet Center in Cambridge, Massachusetts, has 100M observations of NEOs in its database and 27,000 observations are added daily. Today, the NEOO Program has catalogued more than 95 percent of all NEOs over one-kilometer in size and about 25 percent of the 140-meter or larger sized NEO population has been discovered. The current discovery rate of NEOs is approximately 1,000 per year, up 50 percent since 2007. None of the NEOs found to date has a significant chance of hitting Earth in the next century. Thus the near-term risk of an unwarned impact from a large asteroid, and hence the majority of the risk from all NEOs, has been reduced by more than 90 percent.

7. In 1998, NASA commenced an effort with the goal of discovering and tracking over 90 percent of the near-Earth objects larger than one kilometer by the end of 2008. How successful was that effort, what can we learn from it today?

    ANSWER: By the end of 2010, the NEOO Program had reached the goal of cataloguing more than 90 percent of all NEOs over one-kilometer in size at a cost of less than $50M. NASA worked with a number of ground-based observatories and partners as part of our Spaceguard survey to reach that goal; and NASA has now catalogued more than 95 percent of all NEOs over one-kilometer in size.

    Through this process, we have learned that partnerships such as these are essential to meeting the future goals of detecting smaller objects. As such, NASA's NEOO Program has initiated development of several additional capabilities to the NEO detection network, with the recent additional funding it received starting in FY 2012. Some of these involve collaboration on projects with the Defense Advanced Research Projects Agency (DARPA) and the U.S. Air Force, such as background detection of asteroids by the new Space Surveillance Telescope (SST), which is on track to start routinely providing observations in 2013. There is also the planned augmentation of the USAF Panoramic Survey Telescope and Rapid Response System (Pan-STARRS) facility with a second aperture. The wide field of view survey capabilities of these two assets are expected to provide a significant increase in NEO detection rate.

8. When asteroid expeditions are at the center of human spaceflight plans for NASA, and when commercial companies are taking an interest in finding and then profiting from asteroids, and when our ability to avoid a cosmic catastrophe depends absolutely on the knowledge of the orbits of hundreds of thousands (if not millions) of asteroids, why has NASA not funded the effective space-based search telescope needed for all these missions? With Chelyabinsk, will it now receive priority? If not funded, are we to conclude that NASA is not serious about pursuing any of these space endeavors?

ANSWER: NASA leads the world in the detection and characterization of NEOs, and provides critical funding to support the ground-based observatories that are responsible for the discovery of about 98 percent of all known NEOs. However, ground-based telescopes will always be limited to the night sky and by weather. The only way to overcome these impediments is to use the vantage point of space. The privately funded B612 Foundation is planning to build a space observatory called Sentinel that would launch in 2018 and detect 100-meter sized objects and larger that could come near Earth's orbit. Sentinel will employ an infrared telescope from a Venus-orbit that will look out directly opposite the Sun at the space surrounding Earth's orbit in order to see and track near Earth objects. NASA is working collaboratively with the B612 Foundation by providing technical assistance and operational support through a Space Act Agreement. A NASA Technical Consulting Team was established to support the B612 project reviews. NASA will also provide B612 access to our Deep Space Network for telecommunications with the Sentinel spacecraft for commanding and data downlink. NASA is also evaluating reactivating our NEOWISE activity, a very successful use of an Earth-orbiting, Wide-field Infrared Survey Explorer (WISE) space telescope that was used in 2010-2011 to find and physically characterize near-Earth objects.

To find the more numerous smaller asteroids near Earth, NASA's Human Exploration and Operations Mission Directorate (HEOMD) and Science Mission Directorate (SMD), through the Joint Robotic Precursor Activity (JRPA) office, are studying instrument concepts for a mission of opportunity to be hosted on a U.S. government or commercial spacecraft in geosynchronous orbit that would be capable of detecting and tracking asteroids in orbits very similar to Earth's. This modest-sized, wide field telescope would have detectors that operate in the infrared bands where these faint asteroids are more easily detected against the cold background of space. NASA released a Request for Information (RFI) in August 2012 and is studying the instrument concepts that were submitted. Work is also underway to draft an Announcement of Opportunity (AO) to request proposals for Phase A studies. It is likely that NASA could fund up to three instrument concepts for Phase A studies, culminating with a down select to one proposal in FY 2014.

Questions for the Record
Ranking Member Eddie Bernice Johnson
"Threats from Space: A Review of U.S. Government Efforts to Track and Mitigate
Asteroids and Meteors, Part I"
March 19, 2013

1. Will sequestration cause a delay to meeting the Congressionally-mandated goal of detecting 90 percent of NEOs 140 meters in diameter and larger by 2020? If so, by how long?

    ANSWER: The purpose of NASA's Near Earth Object Observations (NEOO) Program is to coordinate NASA-sponsored efforts to detect, track, and characterize potentially hazardous asteroids and comets that could approach the Earth. Currently, sequestration will not have an effect on NEOO Program funding; however, long-term sequestration (past the current fiscal year) could potentially impact programs and projects across the Agency and will be assessed by the Science Mission Directorate as needed.

2. What would be required for NASA to detect 90 percent of NEOs 15 meters in diameter and larger by 2020? What approach(es) would NASA recommend be taken to achieve that goal?

    ANSWER: To find 90 percent of 15 meter NEOs by 2020 would require a program of multiple space-based telescopes to accomplish when it may be several decades before Earth again is hit by something larger than 15 meters. Such a program is currently not achievable with the existing budget profile. The approach within the NEOO Program in the President's FY 2014 budget request is to expand the existing NEO detection and characterization activities to detect the NEO population at a measured pace. Making available more time on existing ground-based observatories, such as the USAF Pan-STARRS, the NASA Infra-Red Telescope Facility (IRTF), or the Space Surveillance Telescope (SST), would be the first step.

    While ground-based surveys are making excellent progress increasing the discovery rate, it would require a space-based infrared NEO telescope to significantly increase the current detection rate. Such a telescope would capture the asteroid's solar energy re-radiated in the infrared and do so without interruptions from daylight and weather. Not only would such a telescope efficiently discover NEOs, it could (unlike ground-based optical discovery telescopes) also estimate their diameters to a confidence level of about 20 percent. The B612 Foundation has announced plans to philanthropically fund an effort to operate an infrared telescope in a Venus-like orbit. NASA has signed a Space Act Agreement with B612 to provide advisory information as well as spacecraft tracking and navigation support. NASA has also funded an advanced infrared detector development that could be employed on an infrared telescope operating at the Sun-Earth L1 position.

In addition, NASA is advancing work on instrument concepts for a mission of opportunity to be hosted on a U.S. government or commercial spacecraft in geosynchronous orbit that would be capable of detecting and tracking asteroids in orbits very similar to Earth's. NASA is also evaluating reactivating our NEOWISE activity, a very successful use of an Earth-orbiting, Wide-field Infrared Survey Explorer (WISE) space telescope that was used in 2010-2011 to find and physically characterize near-Earth objects.

Questions for the Record
Representative Donna F. Edwards
"Threats from Space: A Review of U.S. Government Efforts to Track and Mitigate Asteroids and Meteors, Part 1"
March 19, 2013

1. It is clear that threats from objects 30-50 meters in diameter or smaller, such as the one that unexpectedly entered the atmosphere and exploded over Russia can cause harm.
   - Is NASA's current NEO survey program capable of identifying threats from objects of this size?
   - If not, how important is it to start identifying this class of small threatening objects if we are to come up with an effective protection strategy?
   - How difficult a task would it be?

   ANSWER: Since 1998, NASA has supported several ground-based optical telescope facilities for discovering and following-up NEOs. The progress for finding NEOs larger than one kilometer has been very impressive with a total discovery completion rate of more than 95 percent. However, there are vastly more small NEOs with much larger impact probabilities than large ones. NASA's NEOO Program is capable of identifying these objects; however, the current discovery completion for 30-40 meter sized objects is now less than one percent. While NEOs in this size range can cause local property damage and injuries, our current focus is on completing the Congressionally-mandated survey of NEOs larger than 140 meters, which can cause much more serious damage.

   Small NEOs are difficult to detect in visible light from our current ground-based telescopes because the small, dark objects reflect only a small amount of visible sunlight. In contrast, telescopes sensitive to infrared light detect an object's radiated heat, rather than reflected sunlight; even small, dark asteroids could be detected by a telescope sensitive to infrared light, making these capabilities particularly relevant for future NEO surveys. However, these sensors must operate outside the Earth's atmosphere to be effective.

2. As you know, the Ad-Hoc Task Force on Planetary Defense of the NASA Advisory Council was set up to advise the Council Chairman, you, and NASA Mission Directorates on future agency actions related to Planetary Defense. The Task Force made five recommendations in October 2010 on how NASA should organize, acquire, investigate, prepare for, and lead national and international efforts in Planetary Defense. How were these recommendations subsequently addressed by NASA? What has happened to this Ad-Hoc Task Force since 2010?

   ANSWER: The five recommendations made by the Task Force were to: Organize for Effective Action on Planetary Defense; Acquire Essential Search, Tracking, and Warning Capabilities; Investigate the Nature of the Impact Threat; Prepare the Response to Impact Threats; and Lead U.S. Planetary Defense Efforts in National and International Forums. NASA has acted on these recommendations in several ways.

The purpose of NASA's Near Earth Object Observations (NEOO) Program is to coordinate NASA-sponsored efforts to detect, track, and characterize potentially hazardous asteroids and comets that could approach the Earth. In FY 2010, the NASA budget for the NEOO Program was $5.8M. The final report of the NASA Advisory Council Ad-Hoc Task Force on Planetary Defense was provided to the NASA Administrator in October 2010. NASA's budget formulation process for the subsequent President's budget request resulted in an Administration request of $20M for the NEOO Program in FY 2011 and $20.4M for FY 2012. The FY 2014 budget request for NASA includes $40.5M for the NEOO Program for near-Earth asteroid detection, follow-up and characterization.

Within the increased investment in the NEOO Program, the FY 2014 budget request includes focused support for partnerships and leveraging, including international and commercial partnerships. The objectives of the enhanced NEOO Program are responsive to the Task Force's recommendations to Acquire Essential Search, Tracking, and Warning Capabilities; and Investigate the Nature of the Impact Threat.

In addition, NASA has taken specific action to strengthen the leadership of U.S. planetary defense efforts in national and international forums. The NASA NEOO Program has provided essential leadership to the United Nations Committee on Peaceful Uses of Outer Space action team on the NEO threat. That group is developing a plan for an enhanced international asteroid warning network, impact disaster planning, space-mitigation mission planning should there be a credible threat, and advice on planning and response.

To "Prepare the Response to Impact Threats" on April 3, 2013, the NEOO Program and the Department of Homeland Security's Federal Emergency Management Agency (FEMA) held a one-day simulation of impact disaster response with NEO observation, detection and mitigation community experts and FEMA emergency-response personnel. Also, the Planetary Defense Conference, hosted in April 2013 in Flagstaff, AZ, conducted a half-day international impact-emergency-response exercise with participants from multiple countries.

3. In your prepared statement you noted that "NASA also is investigating development of an instrument that could be hosted on geo-synchronous platforms such as communications, TV broadcast or weather satellites" to detect the more numerous smaller asteroids near Earth. When do you anticipate this instrument being available? How much will it cost and what is your level of confidence that other entities will host the instrument on their platforms?
    - What are the expected outcomes from such a program and what are the criteria for determining whether or not to continue using this approach for NEO detection?

ANSWER: To find the more numerous smaller asteroids near Earth, NASA's Human Exploration and Operations Mission Directorate (HEOMD) and Science Mission Directorate (SMD), through the Joint Robotic Precursor Activity (JRPA) office, are studying instrument concepts for a mission of opportunity to be hosted on a U.S. government or commercial spacecraft in geosynchronous orbit that would be capable of detecting and tracking asteroids in orbits very similar to Earth's. This modest-sized, wide field telescope would have detectors that operate in the infrared bands where these faint asteroids are more easily detected against the cold background of space. NASA released a Request for Information (RFI) in August 2012, and is studying the instrument concepts that were submitted. Work is also underway to draft an Announcement of Opportunity (AO) to request proposals for Phase A studies. It is likely that NASA could fund up to three instrument concepts for Phase A studies, culminating with a down select to one proposal in FY 2014. This effort has the goal of being ready to deploy the first hosted instrument by the end of 2016 for a cost of less than $50M.

4. A NEO object, Apophis, estimated at 325 meters in diameter, has been the focus of much attention and monitoring since it was discovered in 2004. It is projected to have a significant threat of potential impact at some point in the future. What is NASA's current assessment of Apophis' threat and what is needed to improve our understanding of the threat?

    ANSWER: Any significant probability of Apophis impacting the Earth in 2029 was eliminated within a few weeks of its discovery using archived images that allowed a significant extension of the observed orbital track. A small possibility (1 in a few thousand) of impact remained in 2036 until radar observations collected in early 2013 eliminated that event as well. There remains an incrementally small chance of impact in 2068, but it is now assessed at less than 3 in a million.

5. How important is the Arecibo Observatory to NASA's NEO activities?

    ANSWER: The Arecibo Observatory is home to the world's largest and most sensitive single-dish radio telescope, and is one of two radar facilities we use for tracking and improving our knowledge of NEOs. As NEOs are discovered and come into its effective range, the Arecibo Observatory conducts follow-up radar observations to measure specifics such as the distances, sizes and spin rates of the objects, which improve our knowledge of their orbits and help calculate the risks of potential impacts. The rotation rate, shape and reflectivity gathered from the Arecibo images can also give us information about the asteroids' density and surface properties.

QUESTIONS FOR THE RECORD
THE HONORABLE RANDY NEUGEBAUER (R-TX)
U.S. House Committee on Science, Space, and Technology

Threats from Space:
A Review of U.S. Government Efforts to Track and Mitigate Asteroids and Meteors, Part I

1. What does NASA consider an adequate annual funding level for planetary defense activities?

    ANSWER: NASA sponsors a number of activities relating to the search for NEOs and planetary defense related activities under its Near Earth Object Observation (NEOO) program, including work at the international Minor Planet Center (MPC), located at the Harvard-Smithsonian Center for Astrophysics, which collects and correlates NEO orbit data; research at two radio-telescope facilities that help provide precision tracking and characterization of NEOs; surveys conducted by ground-based optical telescopes; and activities at the NASA NEO Program Office at the Jet Propulsion Laboratory (JPL), which coordinates assessments of NEO orbits and impact probabilities. The NEOO program funding was $20.4M in FY 2012. The FY 2014 President's budget request increases this funding to $40.5M, to enhance existing assets that detect and characterize NEOs and initiate development of an instrument that could be hosted on geo-synchronous platforms to detect the more numerous smaller asteroids near Earth.

2. What is the likelihood of such a devastating impact from a near earth asteroid? Are the resources necessary for a credible program worth the cost given the low probability of such an occurrence in our lifetimes?

    ANSWER: Although impact of a large asteroid is an exceeding rare event, one of the key conclusions of the 2003 NASA report entitled "Study to Determine the Feasibility of Extending the Search for Near-Earth Objects to Smaller Limiting Diameters" was that "the benefits derived from all (NEO) search systems match or exceed their costs within the first year of operation." Especially for the larger NEOs, current search efforts are well worth their modest costs, since an early discovery of an Earth threatening NEO would allow the time to safely deflect it with existing technologies. As the search for NEOs continues, and more and more of them are discovered and tracked one hundred or more years into the future, their risks to Earth can be evaluated. More than 95 percent of the largest NEOs (1 kilometer and larger) have already been discovered for a total cost of less than $70M spread over 15 years, and it is reassuring to know that none represent a serious impact threat in the next one hundred years. Despite the low probability of a devastating NEO impact, it seems prudent to continue to invest in strategies for early warning and mitigation.

Question for the Record
Representative Ami Bera

Threats from Space: A Review of U.S. Government Efforts to Track and Mitigate
Asteroids and Meteors, Part 1"

March 19, 2013

1. Astronomy is one of our oldest natural sciences, studied and researched by ancient civilizations before the invention of the telescope, and other modern technologies and sciences. Dating back to 2500 B.C., early records reveal that people kept detailed astronomical accounts of objects they discovered and that practice continues today.

   In the United States, tens of thousands of amateur astronomers create home observatories or assemble at professional observatories and collaborate on finding new celestial objects, ranging from stars, planets, asteroids, etc. These amateur astronomers are found across the globe, even in my home county of Sacranlento, CA as a part of the Astronomy Connection of Sacramento (TAC-SAC) and are making incredible discoveries every day.

   My question is for you General Bolden. What steps can Congress and NASA take to further create and facilitate this open source of information sharing to increase our eyes in the sky for detecting near-Earth objects? How can NASA leverage the passion of amateur astronomy and engage the thousands that practice it to help increase our knowledge and awareness of the asteroids and meteors that are located near Earth?

   ANSWER: Since 1998, NASA has supported several ground-based optical telescope facilities and received observations from numerous amateur astronomers for discovering and following-up NEOs. The international community of NEO researchers is well coordinated and has been working cooperatively for several years. In addition, the international communication and data sharing channels are operating successfully.

   Once a NEO discovery is made, a combination of professional and amateur astronomers provide the critically important follow-up optical observations that allow accurate orbits to be computed and the NEO's motion to then be accurately predicted for more than one hundred years into the future. Radar observations, if available, are especially good for orbit refinement and for determining the NEO's size, shape and rotation characteristics. In addition, many amateur astronomers provide an observed time history of the NEO's ability to reflect light and hence, if these objects are irregularly shaped, these types of observations can be used to determine the rotation rate of the NEO. Given the success of this coordinated effort, NASA will continue to leverage the knowledge and awareness of the NEO community in its entirety.

Appendix II

Additional Material for the Record

SUBMITTED STATEMENT BY REPRESENTATIVE STEVE STOCKMAN, COMMITTEE ON SCIENCE, SPACE AND TECHNOLOGY

Mr. Chairman, thank you very much for focusing Congress' attention on taking effective action on the threats, and solutions to, potentially dangerous meteors and asteroids.

The Chelyabinsk meteor, the flyby of asteroid 2012 DA14, and the 1908 Siberian Tunguska event all offer the dramatic lesson that tracking and mitigating such objects must become a national priority.

We know a large meteor or asteroid could destroy a city and kill millions of people. Unlike in 1908, we now have the ability—and therefore the responsibility—to take effective actions for identifying and avoiding a potentially catastrophic collision.

Under current funding levels, NASA will not be able to meet the Congressional requirement to identify 90% of all objects 140 meters in diameter or larger by 2020. Altering the trajectory of an object in the Earth's path could not be accomplished within decades at current funding levels.

Therefore these objectives must be met with additional and sufficient funds rather than reducing or cancelling funding for existing NASA programs. 'Robbing Peter to pay Paul' would only result in half-hearted efforts which would fail to address the threat from asteroids while at the same time crippling our existing space program. A poorly-funded program will yield poor results.

I am a tireless budget-slasher; however, science, space; and yes, planetary defense are among the few government programs essential to our future.

Advances in technology for planetary defense may provide spinoffs for propulsion to take Americans to Mars and beyond; for cleaning up space debris which threatens satellites and the International Space Station; as well as for more everyday-life applications.

This is of course a worldwide threat, and other nations should participate in developing solutions. However as with all smart space partnerships, it is in our distinct national interest that the United States lead the effort. This will assure that the majority of the technology developed will directly benefit the U.S. economy, and will give the U.S. the ability to block the transfer of our most advanced technology to our potential adversaries. The same technology to track and alter the course of asteroids could have military applications.

The threat from asteroids and meteors is real. America must take the lead to develop practical and effective solutions, reap the technological benefits—lest a decade or two from now we regret our inaction.

LETTER SUBMITTED BY DR. DANTE LAURETTA, DEPARTMENT OF PLANETARY SCIENCES, LUNAR AND PLANETARY LABORATORY

Dante Lauretta
Department of Planetary Sciences
Lunar and Planetary Laboratory

THE UNIVERSITY OF
ARIZONA.
TUCSON ARIZONA

Tucson, Arizona 85721-0092
Phone: (520) 626-1138
Email: lauretta@lpl.arizona.edu

March 6, 2013

The Honorable Lamar Smith, Chairman
House Committee on Science, Space, and Technology
2321 Rayburn House Office Building
Washington, DC 20515

Dear Chairman Smith and Committee Members,

The University of Arizona appreciates the opportunity to discuss the hazards and exploration of near-Earth asteroids. As one of the nation's foremost space sciences universities and leader of both NASA's OSIRIS-REx asteroid sample return mission and the Catalina Sky Survey, a congressionally mandated NASA program to detect potentially hazardous asteroids, we have special expertise in this field.

Asteroids are direct remnants of the formation of the Solar System. They reveal the early Solar System processes responsible for planet formation. Knowledge of their origin and evolution is fundamental to understanding the formation of the Earth and the origin of life. The majority of known asteroids orbit within the main belt between Mars and Jupiter. Near-Earth asteroids are fragments of main-belt asteroids that evolved into an Earth-approaching orbit after a collision event between larger asteroids. Asteroid impacts with the Earth in the past have had a significant role in shaping the history of our planet. As of February 2013, over 9,600 near-Earth asteroids have been discovered.

Catalina Sky Survey

NASA's Near-Earth Objects Observations Program resulted from a 1998 congressional directive to identify 90% of near-Earth objects ≥ 1 km (3,280 feet) in diameter. This effort is known as the *Spaceguard* goal.

The University of Arizona (UA) has led the search for near-Earth asteroids since the early 1980s. Today, the UA Catalina Sky Survey carries out a sustained search for near-Earth asteroids using two telescopes north of Tucson. Catalina has discovered one-third of the known Near-Earth asteroids. In 2008, Catalina made the very first discovery of an asteroid on a collision course with the Earth and was able to alert a worldwide network of observers quickly enough to allow the prediction of its impact area in northern Africa.

Recently, NASA announced that the original *Spaceguard* goal had been achieved, retiring a considerable amount of the total risk from a collision. Despite this milestone, there is general agreement that significant risk remains.

Most near-Earth asteroids less than 1 km remain undiscovered, including an estimated 3,500 objects larger than 100 meters (325 feet) in diameter. Objects of this size are capable of regional destruction, with impact energies equivalent to 300 megatons — about six times larger than the largest thermonuclear device ever exploded. As asteroid sizes decrease, the numbers of objects increase. For comparison, the recent meteor that exploded over Chelyabinsk, Russia was about 20 meters in size and had the equivalent explosive force of 440 kilotons of TNT. Best estimates suggest that over one million Near-Earth asteroids of this size exist. Continued operation of the Catalina Survey will help ensure UA's record of substantial contribution to reducing the threat posed by near-Earth objects.

OSIRIS-REx Asteroid Sample Return Mission

OSIRIS-REx is the United States' premier asteroid mission. It will visit asteroid 1999 RQ36, a carbon- and water-rich object that is also one of the most potentially hazardous near-Earth asteroids. NASA selected this mission for its New Frontiers program in May 2011. The Lunar and Planetary Laboratory at the University of Arizona leads the mission. NASA's Goddard Space Flight Center in Greenbelt, MD is responsible for mission project management. Lockheed Martin Space Systems in Littleton, CO will build and operate the spacecraft. The mission Website is http://osiris-rex.lpl.arizona.edu/.

OSIRIS-REx stands for **O**rigins, **S**pectral **I**nterpretation, **R**esource Identification, and **S**ecurity, **R**egolith **Ex**plorer. OSIRIS-REx will survey asteroid 1999 RQ36 to understand its physical and chemical properties, assess its resource potential, refine the impact hazard, and return a sample of this body to Earth for detailed scientific analysis. This mission is scheduled for launch in 2016 and will rendezvous with the asteroid in 2018. Sample return to Earth occurs in 2023.

OSIRIS-REx is essential to maintain US leadership in near-Earth space in an era of substantial international interest in asteroid exploration. Other nations are actively interested in asteroids as a target for exploration. The Japanese Aerospace and Exploration Agency (JAXA) is developing the Hayabusa 2 sample return mission to asteroid 1999 JU3, the Chinese probe Chang'e-2 has successfully flown by near-Earth asteroid Toutatis, and the European Space Agency is developing a mission to intercept and impact asteroid Didymos in 2022.

Asteroid 1999 RQ36 is a potential Earth impactor. It is a 500-m-diameter asteroid that would enter the Earth's atmosphere with a velocity of 12.9 km/s resulting in impact energy of almost 3,000 megatons. The combined probability of an impact in the late 22nd century (between 2169 – 2199) is 1 in 1,410, one of the highest for any known asteroid (see http://neo.jpl.nasa.gov/risk/a101955.html).

The primary source of uncertainty in assessing the long-term impact probability of asteroids is the Yarkovsky effect, a force that changes the orbit of small rocky objects when they absorb sunlight and then re-emit that energy as heat. OSIRIS-REx will not only investigate the asteroid properties that result in this phenomenon, but also directly measure the Yarkovsky acceleration. OSIRIS-REx thus serves as a "transponder mission," a mission to a potentially hazardous asteroid with the dual objectives of refining the orbit to ascertain whether an impact is impending and characterizing the object to facilitate a possible deflection mission.

OSIRIS-REx is developing critical technologies for exploring near-Earth asteroids including: 1) measurement of the global characteristics of an asteroid, 2) accurate navigation to a specific location on the asteroid surface, 3) successful contact and acquisition of material from that surface, and 4) safe return of the sample to Earth. These operational capabilities are essential as humanity explores near-Earth space to increase our understanding of Solar System bodies and develop *in situ* resource utilization.

A natural extension of the University of Arizona's established experience and knowledge in planetary science, the OSIRIS-REx mission seeks to understand the Solar System scientifically, prepare for human exploration, and assess the risk of one of the most threatening potentially hazardous asteroids.

I'd like to thank Congressman David Schweikert and the US House Committee on Science, Space, and Technology for the opportunity to submit these comments for the record.

Sincerely,

Dr. Dante Lauretta

ADDITIONAL RESPONSES SUBMITTED BY THE HONORABLE CHARLES F. BOLDEN, JR.

Material requested for the record by Rep. Rohrabacher during the March 19, 2013, NEO hearing.

To prepare for the unlikely event where the Earth would be threatened by a collision with a near-Earth object (NEO), we believe an enhanced program would include a steady effort of ground-based observation and monitoring of the detected hazards as they are found (lifecycle cost estimate of up to $600M over 20 years). Further enhancements could include space-based surveys to provide more timely detection of the hazardous population, and technology demonstration missions to test deflection techniques. The costs of these further enhancements are difficult to precisely estimate, but might be on the order of $2.5 – $3B.

Material requested for the record by Rep. Brooks during the March 19, 2013, NEO hearing.

The most important way to prevent the collision of the Earth with a large asteroid is to find the potentially hazardous objects and characterize their orbits as early as possible. NASA's FY 2014 budget request doubles the funding for our Near-Earth Object Observation (NEOO) activities to increase the pace at which we are discovering and characterizing NEOs of all sizes. No known 1-kilometer or larger NEO is an impact hazard to the Earth within at least the next century, and we have evidence that we now know of over 95 percent of them. Even current survey capabilities should allow us to find a 1-kilometer or larger NEO long before it presents any impact threat to the Earth. However, if we assume that a 1-kilometer sized object was discovered and found to be on a collision course with the Earth, there are several factors to consider when estimating the time needed to respond. In short, how quickly we can divert a large asteroid is very much dependent on the scenario.

The three main components of a deflection timetable are: 1) the time required for spacecraft development and launch; 2) the time of flight required to reach the object; and, 3) the time needed to effect the deflection so that the object misses the Earth entirely (which varies based on the technique). Our preliminary estimate is that it would take 1-2 years to build and develop the spacecraft needed for such a mission. The travel time to reach the object will depend on the orbit of the object, the launch vehicle used, and the inclination of the object's orbit with the Earth's orbit; depending on the scenario, the time of flight could be between 1 and 3 years. Finally, the various deflection techniques that could be used (which are spelled out in some detail in the National Research Council's report entitled "Defending Planet Earth: Near-Earth Object Surveys and Hazard Mitigation Strategies: Final Report," especially in Chapter 5) could take an additional 1 to 3 years to have the desired effect on the orbit of the object so that it misses the Earth with an acceptable margin of safety.

Material requested for the record by Rep. Schweikert during the March 19, 2013, NEO hearing.

2012 DA_{14} was discovered on February 23, 2012, by the Observatorio Astronómico de La Sagra (Astronomical Observatory at La Sagra), which uses a half meter-class wide field of view telescope located in the mountains of southern Spain. The telescope is operated remotely by a team of citizen scientists funded in part by ESA's Space Situational Awareness program for both NEO and space debris observations. They could best be described as "semi-pro" astronomers, as they are highly skilled with the detection and astrometric techniques; however, many on the team have other full-time professions. However, it is the NASA-funded professional survey teams (i.e., Catalina Sky Survey, Lincoln Near-Earth Asteroid Research (LINEAR), etc.) that have made more than 98 percent of the NEO discoveries since 1998. Only about 1 percent of discoveries have come from the amateur community since that time, a percentage that is dropping off as we pursue the smaller sized population.

Material requested for the record by Rep. Esty during the March 19, 2013, NEO hearing.

There already exists a very collaborative international network of both professional and amateur astronomers who report all observations on NEOs to the International Astronomical Union sanctioned Minor Planet Center (MPC), hosted by the Smithsonian Astrophysical Observatory at Cambridge, MA, and fully funded by NASA's Near Earth Object Observation Program. This network existed informally for many decades prior to start of NASA's program, but since its inception we have done many things to enhance the operations of the MPC, support dedicated survey teams to greatly increase the detection rate, and created the capability at our Jet Propulsion Laboratory to rapidly provide more precision orbit determination and assess the potential for an impact threat – collective capabilities that are often referred to as the "Spaceguard" network. Also as part of that network a team of astrodynamicists at the University of Pisa, Italy, does parallel analysis of precision orbits so that we can check our answers with them.

The work we have been involved with at the UN Committee on Peaceful Uses of Outer Space in relation to an International Asteroid Warning Network is to encourage more participation in addition to the existing effort from across the member states to enhance the detection and observation capabilities, and to establish more formal reporting and information exchange on any detected hazard so that all nations will have equal access to any knowledge of a potential asteroid threat.

THREATS FROM SPACE: A REVIEW OF PRIVATE SECTOR EFFORTS TO TRACK AND MITIGATE ASTEROIDS AND METEORS, PART II

WEDNESDAY, APRIL 10, 2013

House of Representatives,
Committee on Science, Space, and Technology,
Washington, D.C.

The Committee met, pursuant to call, at 2:00 p.m., in Room 2318 of the Rayburn House Office Building, Hon. Lamar Smith [Chairman of the Committee] presiding.

LAMAR S. SMITH, Texas
CHAIRMAN

EDDIE BERNICE JOHNSON, Texas
RANKING MEMBER

Congress of the United States
House of Representatives
COMMITTEE ON SCIENCE, SPACE, AND TECHNOLOGY
2321 RAYBURN HOUSE OFFICE BUILDING
WASHINGTON, DC 20515-6301
(202) 225-6371
www.science.house.gov

U.S. House of Representatives
Committee on Science, Space, and Technology

Threats from Space:
A Review of U.S. Government Efforts
to Track and Mitigate Asteroids and Meteors, Part 2

Wednesday, April 10, 2013
2:00 p.m. to 4:00 p.m.
2318 Rayburn House Office Building

Witnesses

Dr. Ed Lu, Chairman & CEO, B612 Foundation

Dr. Donald K. Yeomans, Manager, Near-Earth Objects Program Office, Jet Propulsion Laboratory

Dr. Michael F. A'Hearn, Vice-Chair, Committee to Review Near-Earth Object Surveys and Hazard Mitigation Strategies, National Resource Council

U.S. House of Representatives
Committee on Science, Space, and Technology

HEARING CHARTER

Threats from Space, Part II:
A Review of Private Sector Efforts
to Track and Mitigate Asteroids and Meteors

Wednesday, April 10, 2013
2:00p.m. – 4:00 p.m.
2318 Rayburn House Office Building

Purpose

At 2:00 p.m. on April 10, 2013, the Committee on Science, Space, and Technology will hold a hearing titled *Threats from Space, Part II: A Review of Private Sector Efforts to Track and Mitigate Asteroids and Meteors*. This is the second hearing this Congress where the Committee examines the tracking, characterization and mitigation of Near Earth Objects. The hearing will focus on the most viable near-term initiatives within the private sector and the international coordination needed to identify and characterize potentially hazardous near Earth objects.

Witnesses:

- **Dr. Ed Lu,** Chairman & CEO, B612 Foundation

- **Dr. Donald K. Yeomans,** Manager, Near-Earth Objects Program Office, Jet Propulsion Laboratory

- **Dr. Michael F. A'Hearn,** Vice-Chair, Committee to Review Near-Earth Object Surveys and Hazard Mitigation Strategies, National Research Council

Overview

On Friday, February 15, 2013, two events occurred that received worldwide attention. An unforeseen meteor (estimated 50 feet in diameter) exploded in the sky above the Russian city of Chelyabinsk releasing the equivalent of a 300 kiloton bomb, about twenty times the explosive energy of the atomic blast used over the city of Hiroshima. This blast injured nearly 1,200 people and resulted in an estimated $33 million in property damage. On the same day, a small asteroid (150 feet in diameter) discovered by amateur astronomers and tracked closely by NASA passed safely by the Earth, but within the orbital belt of geostationary satellites. Until it entered our atmosphere, the Russian meteor went completely undetected. According to NASA, the two events were unrelated, but raised public awareness of the potential threat from Near Earth Objects (NEOs).

On March 19, 2013, the Committee on Science, Space, and Technology held a hearing titled "Threats from Space: A Review of U.S. Government Efforts to Track and Mitigate Asteroid and Meteors." That hearing addressed the U.S. government's plans and programs to identify, catalog, and coordinate the threat of NEOs.

This hearing will look beyond just the U.S. Government to hear about public-private partnership, commercial private sector, and philanthropic initiatives to survey the sky for asteroids and comets.

Some of the overarching questions:

- Do we have the tools and technology necessary to detect and track Near Earth Objects?
- What are the most viable efforts to focus on in the next 5 to 7 years that will yield the most progress in identifying and cataloging the NEO threat?
- Are we tracking the right size objects, specifically the ones that can cause significant harm?
- Once we identify an object, what are our means of tracking it?
- What are our contingencies and mitigation capabilities if we determine there is a threat from a NEO impact?

The Science, Space, and Technology Committee has been on the forefront of the issues surrounding Near Earth Objects. For example, the NASA Authorization Acts of 2000 and 2005 directed NASA to conduct a survey of the population of NEOs and study mitigation plans. NASA estimates 20,000 potentially hazardous asteroids orbit within the vicinity of the Earth.

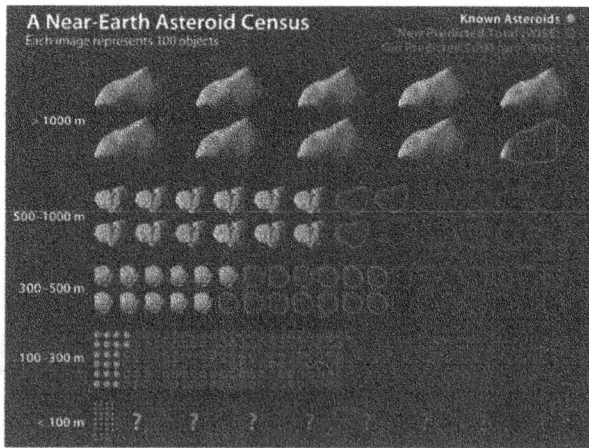

NASA NEO Asteroid Size Model
Credit: NASA/JPL-Caltech

Chairman SMITH. The Committee on Science, Space, and Technology will come to order.

Welcome to today's hearing, which is titled "Threats from Space, Part II: A Review of Private Sector Efforts to Track and Mitigate Asteroids and Meteors." I will recognize myself for an opening statement and then the Ranking Member.

A few weeks ago, our Committee held a hearing to review U.S. Government efforts to track incoming asteroids and meteors. Today, we will follow up by focusing on nongovernmental efforts.

The substantial public interest in this issue indicates the broad fascination with the subject. As witnesses said in our previous hearing, the events of February 15, when an asteroid passed close by the Earth and a meteor struck Russia, were unique in their occurring on the same day. And I am going to hold up a piece of the asteroid that exploded above Russia on February 15. Maybe I ought to take it out of the bag here. Let me—and I am assuming this is not toxic. Is that right? But there it is, a nice size bit of meteorite there, a gift from the Russians. It was given to us by the Principal Investigator of NASA's Asteroid Sample Return Mission, which is slated to launch in 2016.

In our first hearing, testimony about the government's efforts was not reassuring. Most troubling to me was the fact that of the up to 20,000 asteroids that could be labeled as city destroyers, we have identified only 10 percent. And we are unlikely to have the means to detect 90 percent until 2030.

Detecting asteroids should not be the primary mission of NASA. No doubt the private sector will play an important role as well. We must better recognize what the private sector can do to aid our efforts to protect the world.

Today's hearing will help us understand the level of risk, as well as what capabilities we have and those we will need. The President's FY 2014 budget proposal brings necessary attention to this issue in general, but a consensus will have to be reached within Congress before progress can actually be made.

This won't be an effort for one agency, one company, or one country. And in these fiscally challenging times, we can't afford duplication or the inefficient use of our resources. The more we discuss and understand the challenges we face, the easier it will be to facilitate possible solutions.

Now, I will recognize the Ranking Member, the gentlewoman from Texas, Ms. Johnson, for her comments.

[The prepared statement of Mr. Smith follows:]

PREPARED STATEMENT OF LAMAR S. SMITH, CHAIRMAN, HOUSE COMMITTEE ON SCIENCE, SPACE, AND TECHNOLOGY

Good afternoon. A few weeks ago, our Committee held a hearing to review U.S. Government efforts to track incoming asteroids and meteors. Today, we will follow up by focusing on nongovernmental efforts.

The substantial public interest in this issue indicates the broad fascination with this subject. As witnesses said in our previous hearing, the events of February 15, when an asteroid passed close by the Earth and a meteor struck Russia, were unique in their occurring on the same day.

This is a piece of the asteroid that exploded above Russia on Feburary 15th. It was given to me by the Principal Investigator of NASA's asteroid sample return mission, which is slated to launch in 2016.

In our first hearing, testimony about the government's efforts was not reassuring. Most troubling to me was the fact that of the up to 20,000 asteroids that could be labeled as "city destroyers," we have identified only 10%. And we are unlikely to have the means to detect 90% until 2030.

Detecting asteroids should not be the primary mission of NASA. No doubt, the private sector will play an important role as well. We must better recognize what the private sector can do to aid our efforts to protect the world.

Today's hearing will help us understand the level of risk, as well as what capabilities we have and those we will need. The President's FY 14 budget proposal brings necessary attention to this issue in general, but a consensus will have to be reached within Congress before progress can be made.

This won't be an effort of one agency, one company, or one country. And in these fiscally challenging times, we can't afford duplication or the inefficient use of our resources. The more we discuss and understand the challenges we face, the easier it will be to facilitate possible solutions.

Ms. JOHNSON. Good afternoon. I want to join the Chairman in welcoming our witnesses to today's hearing. You each have deep experience and expertise directly related to the hearing topic, and I look forward to your testimony.

As the Chairman has indicated, this hearing is the second that the Committee has held in the opening months of the 113th Congress on the topic of asteroids.

Last month's meteor over Russia and the close passage of a near-Earth asteroid both stimulated public interest in the potential threat posed by asteroids and comets. And this second hearing is certainly a reflection of that interest.

I will not attempt to repeat the sentiments I expressed at our first hearing on this topic and instead will confine myself to a few brief comments.

First, it is clear that from last month's hearing there is still a lot of work to be done to track and characterize asteroids that could potentially impact the Earth and that even relatively small asteroids could do significant damage if they hit in a heavily populated area. So I hope that our witnesses today will help us better understand what will be needed to complete the existing survey, as well as perhaps extend it to a smaller size asteroid.

Second, I want to be one to better understand both the strengths and limits of NASA relying on private organizations such as the B612 for detection of potential Earth-impacting asteroids. My problem is not with the efforts of such organizations to address what they see as an important problem. Instead, my concern is that we have reached a point where our government has to hope that non-governmental organizations will somehow do what the government should be doing but it apparently is unwilling to pay for it. However, if the protection of the planet is not an appropriate role for the Federal Government, I am not sure what is. And finally, before I close, I will note that the President's just-released budget request proposes to invest in a number of asteroid-related initiatives. We will need to closely examine the President's proposals in the coming weeks to fully understand what is being proposed. So I am not going to comment on them in any depth today. Instead, I will simply say that I deeply hope that whatever new initiatives are being proposed will be accomplished accompanied by adequate funding of their own rather than being funded by cannibalizing other important NASA programs. Robbing Peter to pay Paul will not give us sustainable and effective NASA programs. And I hope we will all resist the temptation to do so as we try to address the challenges posed by near-Earth asteroids. Thank you and I yield back.

[The prepared statement of Ms. Johnson follows:]

PREPARED STATEMENT OF RANKING MEMBER EDDIE BERNICE JOHNSON

Good afternoon. I want to join the Chairman in welcoming our witnesses to today's hearing. You each have deep experience and expertise directly related to the hearing topic, and I look forward to your testimony.

As the Chairman has indicated, this hearing is the second that the Committee has held in the opening months of the 113th Congress on the topic of asteroids. Last month's meteor over Russia and the close passage of a near-Earth asteroid have both stimulated public interest in the potential threat posed by asteroids and comets, and this second hearing is certainly a reflection of that interest.

I will not attempt to repeat the sentiments I expressed at our first hearing on this topic and instead wil confine myself to a few brief comments. First, it is clear from last month's hearing that there is still a lot of work to be done to track and characterize asteroids that could potentially impact the Earth. And that even relatively small asteroids could do significant damage if they hit a heavily populated area. So I hope that our witnesses today will help us better understand what will be needed to complete the existing survey as well as perhaps extend it to smaller-sized asteroids.

Second, I want to better understand both the strengths and limits of NASA relying on private organizations such as B612 for detection of potential Earth-impacting asteroids. My problem is not with the efforts of such organizations to address what they see as an important problem. Instead, my concern is that we have reached a point where our government has to hope that nongovernmental organization will somehow do what the government should be doing but is apparently unwilling to pay for. However, if the protection of the planet is not an appropriate role for the Federal Government, I'm not sure what is.

Finally, before I close, I will note that the President's just-released budget request proposes to invest in a number of asteroid-related initiatives. We will need to closely examine the President's proposals in the coming weeks to fully understand what is being proposed, so I'm not going to comment on them in any depth today. Instead, I will simply say that I deeply hope that whatever new initiatives are being proposed will be accompanied by adequate funding of their own rather than be funded by cannibalizing other important NASA programs. Robbing Peter to pay Paul will not give us sustainable and effective NASA programs, and I hope we will all resist the temptation to do so as we try to address the challenge posed by near-Earth asteroids.

Chairman SMITH. Thank you, Ms. Johnson. Other Members' statements will be made a part of the record. And I will introduce our witnesses now.

Our first witness is Dr. Ed Lu. Dr. Lu is the CEO of the B612 Foundation, which aims to build, launch, and operate the Sentinel Space Telescope to help find and track threatening asteroids. He is a former NASA astronaut who flew three space missions and spent six months aboard the International Space Station. From 2007 to 2010, he led the Advanced Projects Group at Google. His teams developed imaging technology for Google Earth Maps, Google Street View, and energy information products, including Google Power Meter. He is also the co-inventor of the gravity tractor, a spacecraft able to controllably alter the orbit of an asteroid. And he has published scientific articles on high-energy astrophysics, solar physics, plasma physics, cosmology, and statistical physics. He holds a bachelor's degree in electrical engineering from Cornell and a Ph.D. in astrophysics from Stanford University.

Our second witness is Dr. Donald Yeomans. Dr. Yeomans is a Senior Research Scientist, Supervisor for the Solar System Dynamics Group, and Manager of NASA's Near-Earth Object Program Office at Jet Propulsion Laboratory in Pasadena, California. His research focuses on the physical and dynamical modeling of comets and asteroids. He was a Radio Science Team Chief for the Near-Earth Asteroid Rendezvous Mission. He has received 15 NASA Achievement Awards and asteroid 2956 was named 2956 Yeomans in honor of his professional achievements. Dr. Yeomans received his Bachelor of Arts degree from Middlebury College and his Ph.D. in astronomy from the University of Maryland.

Our final witness is Dr. Michael A'Hearn. Dr. A'Hearn is a Professor in the Astronomy Department at the University of Maryland. He is the Principal Investigator for the Deep Impact Mission and NASA's Discovery Impact Mission in NASA's Discovery Program and for the Small Bodies Node of NASA's Planetary Data System. His research emphasizes the study of comets and asteroids. Dr. A'Hearn received a Bachelor of Science degree in physics from Boston College and a Ph.D. in astronomy from the University of Wisconsin, Madison.

Now, we welcome you all. Thank you for being here. And Dr. Lu, we will begin with you.

STATEMENT OF DR. ED LU, CHAIRMAN AND CEO, B612 FOUNDATION

Dr. Lu. Thank you, Members of the Committee, and thank you, Chairman Smith, especially for your leadership on this issue.

So my name is Ed Lu, and I am CEO of the B612 Foundation. I want to thank you for the opportunity to testify before the Committee to describe the B612 Foundation and its Sentinel Space Telescope project and the importance of that project.

The B612 Foundation is a Silicon Valley-based nonprofit that is building, launching, and operating the Sentinel Space Telescope, which will find and track threatening asteroids. So NASA, at the direction of Congress, has found and tracked 95 percent of the large asteroids, those larger than a kilometer, that would likely end civilization were they to hit. So they have done a great job on that. And none of these civilization-enders is known—thus far discovered—is known to be on an impact course anytime in this upcoming century. So that is the good news.

But NASA has not even come close to finding and tracking the one million smaller asteroids that might only just wipe out a city or perhaps collapse a rural economy if they hit in the wrong place. I would like to clarify something, and so I thought this image might be of help. I just show here a football stadium, which I understand now is Heinz Field in Pittsburgh, and we show a couple of asteroids there, but just for scale.

A 140-meter asteroid is not shown here, but it would roughly fit inside that stadium. And that is the size—when they hit, that would release about 100 megatons of energy, which is roughly five times all the munitions used in World War II. Okay. So that is much larger than a city killer. That is a regional killer. Okay. And NASA discovered and observed, tracked less than 10 percent of the asteroids in that size range, sort of the stadium-sized ones.

A 40-meter asteroid, which is the larger of these two, is what you would really call a city killer. The last one to hit was in 1908 in Tunguska, and that had an impact energy about three to five megatons of energy. It destroyed about 1,000 square miles of Siberian forest. And we have observed and tracked well less than one percent of the million or so asteroids of that size. So if you ask how many city killers out there have we found and have tracked? Less than one percent is the answer. And there is about a 30 percent chance that there will be another impact of a city killer sometime this century, somewhere on the surface of the Earth. Just for reference, the smaller one shown there is about the size of the one that struck Chelyabinsk last month.

So we simply don't know when the next catastrophic asteroid impact is going to be, because we simply haven't yet tracked the great majority of asteroids. Again, less than one percent of these city killers have been tracked. Yet we have the technology to deflect asteroids, and Dr. A'Hearn will probably talk a little bit about Deep Impact. It is—you—which is an experiment to actually hit an asteroid with a small spacecraft, and that is all you really need to do in most cases if you find the asteroids well in advance and—because you can't deflect an asteroid that you haven't yet tracked. Our technology is useless against something we haven't found.

So that is why our number one priority from the standpoint of planetary defense is to find and track asteroids as soon as practical. You can't deflect an asteroid you haven't yet found, or for that matter, you can't capture it, you can't visit it, you can't mine it, you can't explore it until you have found it.

So finding and tracking the roughly one million or so city killer asteroids in a reasonable time frame requires a system that can find tens to hundreds of thousands of them per year, right? If you are going to get to a million, you need to find them at a very high rate. Anything less than that, from a planetary defense standpoint, is just playing around the edges.

So this task of finding those smaller asteroids cannot be done even by large ground-based telescopes, optical telescopes, and it especially cannot be done by small telescopes. So—and that is because asteroids are not only small but they are dark. Their color is often as dark as charcoal, and that makes them really dim. So these smaller asteroids are only spotted currently when they come very, very close to the Earth. So, because most of the large asteroids have been found, unfortunately, that means that amateur astronomers and people with smaller telescopes can no longer substantially contribute to this particular effort, nor will small space-based optical telescopes such as have been proposed by some commercial companies, they will not make a dent in this problem.

But the fact that asteroids are dark can be used to our advantage, because when they are small and dark, they absorb light from the sun and they are warmed. And that means they are brighter than the background sky if you observe them in infrared. And when you observe them in infrared, you can see them at much greater distances than you can with optical telescopes.

So as described in the National Academies report "Defending Planet Earth," if you want to find a substantial fraction of city killer asteroids, you need a space-based infrared telescope. So that is what the B612 Foundation is doing. Our Sentinel Space Telescope is going to launch in 2018. It will orbit the sun about 30 million miles closer to the sun than the Earth in a solar orbit that is similar to the orbit of Venus, and that means Sentinel will not have a blind spot because—like Earth-based telescopes, which can only look at night looking away from the sun. Sentinel will always look away from the sun, looking outwards at Earth's orbit.

So it will find and track as many asteroids as have been discovered by all other telescopes combined just in the first month of operation. Over six and a half years it will find over half a million asteroids, including more than 90 percent of the sort of stadium-sized ones, the regional killers, and the majority of those that are just city killers, the larger of these two asteroids. These asteroids will be tracked accurately enough to know if any of them is going to be on a course to hit Earth this century.

So to carry out this mission, the B612 Foundation has assembled perhaps the finest technical team I have had the privilege of working with in my nearly two decades of involvement in aerospace, including 12 years as a NASA astronaut. The fact that we were able to recruit such a team is, I think, a testament to the inspiring and urgent nature of this mission. As we tell these people, who wouldn't want to have a chance to save the world? And that is really what I think drew them to the mission.

So our major partner in transmitting our data back, as well as allowing some NASA experts to sit on some of our technical review panels, including, for instance, Dr. Yeomans here. The data generated by Sentinel will not only protect the people of planet Earth but will form the basis of future exploration and scientific missions.

So a unique aspect of B612 is that we are carrying out this mission as a nonprofit. We do not receive any government financial support, and we are relying upon donations from individuals and foundations. These donors understand the importance of cataloging the environment we inhabit and the solar system, and they as individuals are making Sentinel happen because they know that our future may depend upon it.

So make no mistake, raising this amount of money philanthropically with no expectation of financial return from our donors is challenging. But being a nonprofit has forced us to be very focused, and I believe it has made us resourceful. Our progress has been swift and we are approaching now the second of our eight milestones leading up to launch.

The B612 Foundation is managing this project in an innovative Silicon Valley fashion with the rigor of a NASA project. So we are able to carry out this mission at what we believe to be about 60 percent of the cost as if it had been procured via federal procurement.

So I should point out that the core technologies that Sentinel uses that allow us to detect dark objects via their infrared admissions would be useful to a number of federal agencies, including NASA, and there may be an opportunity to expand our existing public-private partnership with NASA in a manner that leverages our private donations, accelerates our technical progress and, in the end, provides the data that could protect us all.

So we can protect the Earth from asteroid impacts, but we can't do it if we don't know where those asteroids are. And that is why the Sentinel telescope is so important.

Chairman SMITH. Okay, Dr.——

Dr. Lu. I can't think of a more inspiring mission. Thank you.
[The prepared statement of Dr. Lu follows:]

Testimony to the U.S. House of Representatives Committee on Science, Space and Technology

Hearing on Threats from Space: A Review of Private and International Efforts to Track and Mitigate Asteroids and Meteors

April 10, 2013

Dr. Edward T. Lu

CEO - B612 Foundation

My name is Ed Lu, and I am the CEO of the B612 Foundation. Thank you for the opportunity to testify before the House Science, Space and Technology Committee to describe the B612 Foundation Sentinel Space Telescope project and its importance. The B612 Foundation is a Silicon Valley based nonprofit that is building, launching, and operating the Sentinel Space Telescope, which will find and track threatening asteroids.

NASA's Spaceguard Survey has already discovered more than 90 percent of asteroids larger than 1km. Why must we identify and track asteroids smaller than this?

The impact of a 1 km or larger asteroid would have energy of 40 thousand megatons, and would likely end human civilization regardless of where on the Earth it occurs. However smaller yet still potentially catastrophic asteroids are still largely not tracked. For example, the impact of a 140 meter asteroid would release several times more energy than all the munitions used in WWII. Yet have only observed and tracked less than 10% of asteroids in this size range. The impact of a 40 meter asteroid "city killer" such as the one that struck Tunguska on June 30, 1908 obliterated nearly 1000 square miles. Yet we have only observed and tracked less than 1% of asteroids in this size range. We have the technology to deflect asteroids to prevent an impact on Earth, but this technology is useless until we find asteroids first. We cannot deflect (or for that matter capture, visit, or explore) an asteroid that we haven't yet found. We simply do not know when the next catastrophic asteroid impact will be, because we have not yet tracked the great majority of asteroids.

Why is an infrared space telescope needed to discover and track these smaller yet still dangerous asteroids?

Asteroids are not only small but are often as dark as charcoal. That makes asteroids difficult to spot with ground based optical telescopes because not only are they dim, but the background sky is bright. But the fact that asteroids are dark can be used to our advantage if we observe in infrared. Asteroids are much easier to detect in infrared because they are warmer and thus brighter than the background sky in these wavelengths, and can therefore be seen at much greater range using infrared as opposed to optical. Such infrared observations can only be made from space, as the Earth's atmosphere absorbs these infrared wavelengths of light. The National Academies Report "Defending Planet Earth" published in 2010 describes how finding a substantial fraction of "city killer" asteroids like the 1908 Tunguska asteroid will require a space based infrared telescope.

Finding and tracking the roughly 1 million asteroids of this size in a reasonable timeframe requires a system capable of finding tens to hundreds of thousands of asteroids per year. This cannot be done even by large ground based optical telescopes, and it especially cannot be done by small optical telescopes. That means amateur astronomers unfortunately will not substantially contribute to this effort, and neither will small space based optical telescopes which some commercial companies have proposed to operate. Such a task requires an infrared space telescope.

Why is it critical to place Sentinel in Solar orbit (similar to the planet Venus)?

Asteroids that will hit Earth have orbits that cross Earth's orbit, and therefore are sometimes located in the direction of the Sun when viewed from Earth. Earth based telescopes cannot observe these asteroids when they are located in their "blind spot", i.e. when they are interior to the Earth. However, Sentinel will orbit the Sun interior to the Earth, in a solar orbit similar to that of the planet Venus. From that vantage point, Sentinel will be able to continuously look outwards away from the Sun while scanning Earth's orbit. This vantage point combined with Sentinel's ability to track asteroids from greater distances, means that Sentinel will typically be able to track an individual asteroid for several months at a time, which allows the orbit of that asteroid to be determined accurately. This is critical because many asteroids will have orbits which at first may appear to pose a threat to Earth until further observations can be used to refine our knowledge of the asteroid orbit well enough to rule out an impact. This is problem for telescopes located on or near Earth, as many asteroids can only be observed for a few weeks and then cannot be observed for long periods of time (often many years) because these asteroids recede in their orbits to the other side of the Sun for extended periods. Sentinel will orbit the Sun every 8 months, and so it will be able to observed and track these asteroids much more frequently, and therefore will be able to refine the orbits of such asteroids much faster. This will reduce incidences of asteroids having long periods of uncertainty such as we witnessed for the asteroid Apophis from 2004 until about 2010 (when our data was insufficient to be able to rule out an impact with Earth).

What is the role of the B612 Foundation and what makes it unique?

The B612 Foundation is a Silicon Valley based nonprofit that is philanthropically funded. We are funding, building, launching and operating the Sentinel Space Telescope. Because the problem of asteroid impacts is inherently a worldwide problem, the B612 Foundation has donors and supporters from around the world.

Our prime contractor is Ball Aerospace, located in Boulder, CO. Ball has previously built the Kepler Space Telescope, and the Spitzer Infrared Space Telescope on which Sentinel is largely based. We do have some non-financial support from NASA, which is providing use of the antennas of the Deep Space Network for telemetry and tracking, in addition to some technical consulting.

One of the unique aspects of the Sentinel mission is the way it is being managed. We are procuring Sentinel under commercial fixed price terms, much like communications and Earth observing spacecraft are procured. This is the first interplanetary mission to be managed in this fashion. We believe this is possible because Ball Aerospace has substantial experience with similar missions and much of Sentinel is based on that hardware, because there are no fundamentally new technologies which must be developed, and because both B612 and Ball have assembled world class technical teams. The B612 Foundation is managing this project in an innovative Silicon Valley fashion but with the rigor of a NASA project.

What is the status of the Sentinel Space Telescope project?

Sentinel is planned to launch in July 2018. The technical and management teams at both B612 and Ball are largely in place. There are 8 major milestones between project inception and launch. The first major milestone, the Concept and Implementation Review, was completed in September of 2012. The next major milestone, the Systems Definition Review, is scheduled for late fall of 2013. Prototype infrared imaging detectors have been fabricated and are currently in test.

Dr. Ed Lu, Chairman and CEO

Ed Lu is a former NASA astronaut who flew three space missions including 6 months on the International Space Station.

From 2007-2010, he led the Advanced Projects group at Google, where his teams developed imaging technology for Google Earth/Maps, Google Street View, and energy information products including Google PowerMeter.

He is the co-inventor of the Gravity Tractor, a spacecraft able to controllably alter the orbit of an asteroid. He has published scientific articles on high-energy astrophysics, solar physics, plasma physics, cosmology, and statistical physics. He holds a PhD in astrophysics from Stanford University, and a Bachelors degree in Electrical Engineering from Cornell University.

Chairman SMITH. You are at nine minutes and we need to move on. Are you—can you conclude?

Dr. LU. That was my conclusion.

Chairman SMITH. Okay. Good timing.

Dr. Yeomans.

STATEMENT OF DR. DONALD K. YEOMANS, MANAGER, NEAR-EARTH OBJECTS PROGRAM OFFICE, JET PROPULSION LABORATORY

Dr. YEOMANS. Mr. Chairman and Members of the Committee, thank you for the opportunity to discuss some issues related to near-Earth objects, and thank you all for your continuing interest in this topic.

As noted by the Chairman, back on February 15, Friday, we had a 40 meter-sized object that passed within 17,200 miles of the Earth's surface and passed 5,000 miles within the geosynchronous ring of communication satellites that were announcing its arrival. Sixteen hours earlier on the same day, we had an impact over Chelyabinsk, Russia, of an 18 meter-sized object coming in at 42,000 miles per hour weighing 11,000 tons. And although I have been upstaged by Chairman Smith, I also have a piece of the rock that you may want to look at after the hearing.

My point is that the close approach was a 1-in-40 year event for an object of this size getting that close. The impact of the smaller object over Chelyabinsk is a 1-in-100 year event, so very unlikely events do happen sometimes on the same day within 16 hours. Asteroid impacts with the Earth are extremely unlikely, but they could cause global problems. But if we discover them early enough, we have the technology to deflect them.

Significant progress has been made to discover and understand the physical characteristics of near-Earth asteroids, largely as a result of NASA-supported efforts. For example, as pointed out, over 90 percent of those near-Earth asteroids larger than a kilometer have been found, and we have integrated their motions for 100 years into the future, and none of them represent a threat. About 25 percent of 140 meter-sized objects have already been found, and likewise, they do not represent a threat.

So the goal is to find and track 90 percent of the 140 and larger sized objects, and in so doing, we will reduce the threat of all objects of all sizes to a 99 percent level.

A thousand new near-Earth asteroids are discovered each year, almost all of them as a result of NASA-supported surveys. Twenty-seven thousand new asteroid observations per day are added to the archives at the Minor Planet Center in Cambridge, Massachusetts, and there is an increasing pace with which observations of near-Earth asteroid physical characteristics are being taken, including optical measurements, near infrared measurements, and radar measurements.

The vast majority of near-Earth asteroid discoveries are currently being made by the Catalina Sky Survey near Tucson, Arizona, the Pan-STARRS Survey in Hawaii, and the Linear Program near Socorro, New Mexico. And these surveys are continuously improving their discovery efficiencies, and the next generation of near-Earth asteroid survey telescopes and cameras are under development.

However, as pointed out by Ed, still undiscovered are 50 to 100 of the largest near-Earth asteroids and several thousand near-Earth asteroids larger than 140 meters. In fact, there was a two-kilometer—a new two-kilometer-sized asteroid that was announced today, so we still have a handful of large ones to find and several thousand of the smaller ones that are 140 meters and larger.

A dramatic increase in the near-Earth asteroid discovery efficiencies is achievable using space-based infrared telescopes, either in a Venus-like orbit, as pointed out by Ed, or located about a million miles on the sunward side of the Earth at the so-called L1 point. The goal is to find the large near-Earth asteroids early enough to mount a deflection mission if necessary. The easiest and fastest deflection technique involves impacting a spacecraft on the asteroid with a rendezvous spacecraft there to monitor the success and verify that the object was moved just enough so that in 10 or 20 years, when it was predicted to hit the Earth, it would miss by a wide margin.

What about the undiscovered millions of small near-Earth asteroids larger than 30 meters that are most likely to hit the Earth, the city killers, as Ed pointed out? Finding most of these near-Earth asteroids would be extremely challenging. Perhaps a cost-benefit study could establish the appropriate threat levels where it would make more sense to simply warn of an asteroid impact rather than finding it early enough to mount a deflection campaign.

NASA is currently supporting a program called ATLAS at the University of Hawaii that is designed to find small objects a few days or a few weeks prior to impact. And the objective there, of course, is civil defense. If you find it several days in advance, you could evacuate if the object was threatening a populated area.

So, in summary, with the current near-Earth asteroid threat identification process in place, and with considerable augmentations to NASA's Near-Earth Object Observation Program, we can determine which near-Earth objects represent potential future threats and do so with enough time to either deflect the larger objects or warn of the arrival of the smaller ones.

Thank you for your attention.

[The prepared statement of Dr. Yeomans follows:]

HOLD FOR RELEASE
UNTIL PRESENTED
BY WITNESS
April 10, 2013

Statement of
Donald K. Yeomans
Manager, NASA Near-Earth Object Program Office
Jet Propulsion Laboratory

before the

Committee on Science, Space, and Technology
U.S. House of Representatives

Mr. Chairman and members of the Committee, thank you for the opportunity to appear today to discuss the progress and plans being made to discover, track and characterize the population of near-Earth objects that may pose threats to Earth.

The Importance of Near-Earth Objects: Near-Earth objects, commonly called NEOs, are comets and asteroids that can pass within about 28 million miles of the Earth's orbit. While icy active comets may occasionally pass close to Earth, it is the difficult-to-find, but far more numerous asteroids that are of most concern in near-Earth space today. Near-Earth objects are scientifically important because they represent the bits and pieces left over from the solar system formation process. Collisions with the early Earth likely brought much of the water and carbon-based materials that were the building blocks of life. Once life did form, subsequent collisions punctuated evolution, allowing only the most adaptable species to evolve further. We humans likely owe our origins and current position atop the food chain to these near-Earth objects.

While the vast population of near-Earth objects is a relatively recent discovery, they are of utmost importance in the study of the solar system's origin and our own origins, and they will likely play a major role in the future, providing building materials, water and fuel resources for interplanetary exploration and development. It is ironic that the near-Earth objects that are the easiest to reach for robotic or human exploration, and the easiest to exploit for their mineral and material wealth, are the same objects that represent the most serious potential threats to Earth. While finding them is important for future space resource development, we also need to find them - before they find us.

A Recent Hit and a Miss: Less than two months ago, on Friday, February 15, 2013, a 60-foot sized asteroid, traveling at 42,000 miles per hour, entered the Earth's atmosphere near Chelyabinsk in central Russia. It was heated and violently compressed by the atmospheric pressure and exploded about 14 miles above the surface, producing a descending shock wave with an energy of approximately 440 thousand tons of TNT explosives. Most of the asteroid itself was reduced largely to dust, but it also produced thousands of small fragments which fell to the ground as meteorites. Over 1,200 people were injured by the effects of the shock wave – mostly from broken glass. Given the millions of similarly sized objects in Earth's neighborhood, a collision by one would be expected about once every 100 years on average. Coincidently, only 16 hours later, a larger 130-foot asteroid called 2012 DA14 that we had been tracking for a year came from a different direction, passing within 17,200 miles of the Earth's surface, 5,000 miles within the ring of communications satellites broadcasting the news of its arrival. An approach this close to Earth's surface by an object of this size is expected to occur every 40 years or so, on average. So, on the same day, we witnessed a once in a 100-year event and an unrelated once in a 40-year event. Here we have a nice example for science teachers to show that even extremely unlikely events in nature do happen – even two on the same day.

Because it was found a year in advance, we were able to accurately predict the close Earth passage of asteroid 2012 DA14 on February 15, and we knew that it would not hit the Earth. However, the small asteroid that impacted the Earth's atmosphere over Russia arrived unannounced because it came from the direction of the Sun, and was hence unobservable with Earth-based telescopes. Discovering and identifying relatively small Earth impactors among the millions of asteroids in the Earth's neighborhood represents a significant challenge. Because there are so many more smaller asteroids than larger ones, the smaller ones hit the Earth's atmosphere more frequently. There are about ten million 20-meter sized asteroids like the one that exploded over central Russia two months ago, and their frequency of collision with the Earth is about once every 100 years, on average. When these small ones do hit, we expect them to break up in the atmosphere and cause only localized damage on the ground. Asteroids larger than one kilometer, on the other hand, could not only penetrate the atmosphere and impact the Earth's surface, they could also cause ejecta clouds that can affect weather patterns, produce firestorms and acid rain, and seriously harm global society and economics. The number of such large asteroids, however, is far smaller, only about 1000, and the frequency with which they impact the Earth is much lower, only about once every 700,000 years on average. But, the events of February 15, 2013, demonstrate that even extremely improbable events can happen, and that it is prudent to pay attention to the problem of finding and tracking all potentially hazardous near-Earth asteroids. And the focus should not be restricted to just the large near-Earth asteroids; mid-sized objects in the 100 to 500 meter range also pose a serious risk, since they could devastate an entire regional area.

The Spaceguard Goal: The population of near-Earth objects one kilometer and larger appropriately received the most attention in the early years of NASA's Near-Earth Object Observations program. In May 1998, a NASA representative announced plans to the House Subcommittee on Space and Aeronautics: NASA would find and track at least 90% of all the near-Earth objects larger than one kilometer; this became known as the "Spaceguard" goal. In December 2005, President Bush signed into law the George E.

Brown, Jr. Near-Earth Object Survey Act, that gave NASA a broader and more ambitious goal, to detect and track at least 90% of the near-Earth objects larger than 140 meters in diameter, and to characterize the physical properties of a representative sample of this population.

Significant Progress Has Been Made: When I last had the honor to address this Committee in November 2007, about 80% of the NEOs one kilometer or larger had been discovered and only a few percent of the smaller 140 meter objects. Today, the Spaceguard goal of discovering 90% of the large NEOs has been exceeded and about 25% of the 140 meter or larger sized NEO population has been discovered. Today, the discovery rate of NEOs is about 1000 per year, up 50% since 2007. The Minor Planet Center in Cambridge, Massachusetts, has 100 million observations of NEOs in its database and 27,000 observations are added daily. Fully 96% of all NEOs were discovered by NASA-funded surveys. The vast majority of all current NEO discoveries are being made by the Catalina Sky Survey, operating near Tucson, Arizona, the Pan-STARRS survey operation atop Haleakala, Maui, Hawaii, and the LINEAR survey near Socorro, New Mexico.

None of the NEOs found to date have more than a tiny chance of hitting Earth in the next century. Thus the near-term risk of an unwarned impact from large asteroids, and hence the majority of the risk from all NEOs, has been reduced by more than 90%. Assuming none are found to be an impact threat, discovering 90% of the 140 meter sized objects will further reduce the total risk to the 99% level. By finding these objects early enough and tracking their motions over the next 100 years, even those rare objects that might be found threatening could be deflected using existing technologies. For example, a spacecraft could purposely ram the asteroid, modifying its orbital velocity by a very small amount, so that over several years its trajectory would be modified and its predicted impact of Earth in the future avoided by a safe margin. The autonomous spacecraft navigation required to effect such a collision was successfully demonstrated in July 2005 when NASA's Deep Impact spacecraft purposely rammed comet Tempel 1 to better understand the comet's structure and composition.

There have also been dramatic increases in the rate with which observations have been made to understand the physical nature of these NEOs, their so-called "characterization". These include infrared observations that are used to infer asteroid sizes and compositions and radar observations that are used to determine asteroid sizes, shapes, rotation rates and whether or not an asteroid has a moon. For example, since 2007 when I last addressed this Committee, there has been a 250% increase in the number of infrared observations of NEOs made at the NASA supported Infrared Telescope Facility in Hawaii. During 2012 alone, the number of radar detections of NEOs at both the Goldstone facility in California and the Arecibo facility in Puerto Rico has tripled compared to the average of the previous ten years. Largely as a result of NASA support, there has been extraordinary progress in the last six years for the discovery and physical characterization of NEOs.

There has also been significant progress within the NEO Action Team associated with the UN Committee on the Peaceful Uses of Outer Space (COPUOS) to encourage and integrate more international efforts for NEO detections, for addressing deflection issues and for establishing protocols to be used by the international community in response to a potential NEO threat.

Possible Next Steps For Ground-Based Systems: There is still much work to be done. About 50-100 NEOs larger than one kilometer remain undiscovered, along with about 13,000 NEOs larger than 140 meters and millions of objects larger than about 30 meters in extent - the approximate minimum size for a common stony asteroid to cause significant ground damage.

It is important that the current NEO discovery surveys, operating with one to two meter-class optical telescopes continue their nightly searches and continue to improve their equipment, operations and data processing. These ground-based optical telescopes will continue to significantly add to the count of large NEOs discovered, but the current search assets will not be able to reach the goal of finding nearly all of the population of 140-meter sized objects within a reasonable time period because of their limited capability to detect these very dim objects.

The existing Pan-STARRS1 (PS1) system operates a 1.8-meter aperture telescope on the island of Maui but this instrument only focuses its attention on NEO observations for about 11% of its observing time because of other science objectives. Even so, PS1 currently provides about 25% of the NEO discoveries, second only to the Catalina Sky Survey. Suitable funding to increase the percentage of time devoted to NEO searches on Pan-STARRS1, at the expense of other science, would accelerate the current NEO discovery rate, as would the full time or part time use of a second Pan-STARRS2 telescope that is nearing completion adjacent to the Pan-STARRS1 facility on Maui.

An important planned future contributor is the Space Surveillance Telescope (SST), a 3.5-meter wide-field telescope that is being developed by MIT's Lincoln Laboratory for DARPA and the US Air Force. When fully operational in late 2014, this telescope will scan a wide region centered on the equatorial band of the night-time sky. Investigations are ongoing to better understand the efficiency with which this telescope will discover NEOs and what sort of scheduling might be intermingled with its prime mission of man-made space object surveillance to carry out these NEO observations.

The most effective, ground-based NEO detection telescope that is currently in planned development is the Large Synoptic Survey Telescope (LSST), a 8.4-meter aperture, wide-field telescope that is planned to begin operations in Chile in the early 2020s. To be funded by the National Science Foundation and a consortia of private and international agencies and universities for a variety of science programs, simulations have suggested that the shared use of LSST could catalog approximately 25% of the 140 meter sized NEOs within 5 years of operations and about 45% in ten years.

The View From Space: Especially for the population of undiscovered sub-kilometer sized objects, space-based infrared telescopes would be a more efficient discovery system than the current ground-based optical surveys. This is because asteroids emit considerable heat, not just reflected sunlight, and this heat makes them bright in the infrared wavelengths, but these wavelengths are also unfortunately heavily filtered by the Earth's atmosphere. In addition, the view from an observatory orbiting the Sun interior to the Earth's orbit would have far better viewing coverage of hazardous objects farther away from Earth. Furthermore, a space-based telescope would not have to deal with downtime due to weather and daylight. Ground-based telescopes have difficulty distinguishing a large, dark asteroid from a small, bright asteroid, often making asteroid size measurements very uncertain. On the other hand, space-based infrared measurements can infer an asteroid's size with an uncertainty of only about 10% and its reflectivity to about 20%. These types of measurements were demonstrated in 2010 when the highly successful NEOWISE effort mined asteroid discoveries, sizes and reflectivities in the data produced by the Wide-field Infrared Survey Explorer (WISE) satellite.

If the goal is to complete the survey of 140 meter sized objects more quickly, the 2010 National Research Council report entitled "Defending Planet Earth" indicated that a space-based infrared telescope in either a Venus-like orbit or interior to the Earth on the Sun-Earth line (L1 point) would be far more efficient finding NEOs than would existing, or planned, ground-based optical surveys. For the more numerous population of smaller NEOs that can still do significant ground damage, an infrared telescope at L1 would be well positioned to find those smaller objects making close Earth approaches. A successful space-based IR survey telescope in a Venus-like orbit would be very effective in discovering NEOs further in advance and providing positional observations unavailable from Earth-based telescopes. Together these observations would allow a faster refinement of an asteroid's orbit so that impact predictions could also be updated more quickly. Hence these space-based observations might provide an early "all clear" and avoid otherwise unnecessary concern and unneeded deflection mission planning or initiation.

Threat Mitigation vs. Threat Warnings: For the millions of small NEOs, in the range of 30 to 50 meters, it would be extremely challenging to find the majority of this population far enough in advance to first determine which ones represent a threat and then deflect them safely away from Earth. And meeting such a challenge may not be cost effective. It may be sufficient to simply detect these small asteroids a few days or weeks prior to Earth impact so that appropriate warnings could be made and evacuations undertaken similar to hurricane emergencies in the unlikely case where populated areas of Earth would be threatened. A warning of this type would also assure affected nations that the coming explosive blast would be a natural phenomena rather than a hostile act.

The NASA-funded ground-based ATLAS system currently under development at the University of Hawaii is a relatively low cost, wide-field telescopic survey designed to patrol the entire accessible night sky every night to provide suitable impact warnings for small asteroids on near-term Earth impacting trajectories. Simulations suggest that the ATLAS system, consisting of 3 to 4 sites worldwide, will find almost all objects larger than 30 meters coming at us from the night sky and provide a week's warning time. Current search programs are designed to find larger potentially hazardous objects well in advance of a predicted impact so that existing technologies (e.g., spacecraft rendezvous and impacts) could be employed to deflect the object out of harm's way. One of the issues with which policy makers will need to wrestle is where to draw the line as to the minimum NEO size that represents so large a threat as to require deflection attempts. Objects below that limit would then require only advance warning. Cost benefit studies would shed some light on this issue.

Summary: The NASA-supported NEO observations program is proceeding extremely well, and the rate with which NEOs are being discovered and physically characterized is increasing each year. There are viable options for accelerating the current NEO search efficiencies either using next-generation, ground-based optical surveys or the even more efficient space-based infrared surveys. The use of both ground-based and space-based assets would be the most effective option for quickly finding 90% of the NEO population larger than 140 meters. Robust future NEO search programs and the attendant physical characterization efforts could provide a large number of target bodies for scientific study, for future robotic and human exploration and for future resource development. These same surveys could also identify which of the discovered NEOs represent potential future threats and do so with enough time to either deflect the object, or warn of its arrival.

Donald K. Yeomans

At the Jet Propulsion Laboratory, Don Yeomans is a JPL Fellow, Senior Research Scientist, Manager of NASA's Near-Earth Object Program Office and Supervisor of the Solar System Dynamics Group.

Dr. Yeomans was the Radio Science Team Chief for NASA's Near-Earth Asteroid Rendezvous mission and the NASA Project Scientist for the Japanese mission that landed upon, and returned a sample from, a near-Earth asteroid. He was also a scientific investigator on NASA's Deep Impact mission that successfully impacted comet Tempel 1 in July 2005.

He provided the accurate predictions that led to the recovery of comet Halley at Palomar Observatory on October 16, 1982 and allowed the discovery of 164 BC Babylonian observations of comet Halley on clay tablets in the British Museum. His group at JPL is responsible for providing predictions for future close Earth approaches and impacts by comets and asteroids.

Dr. Yeomans has received 19 NASA Achievement Awards including an Exceptional Service medal and a Distinguished Service Medal – NASA's highest award. Asteroid "2956" was renamed asteroid "2956 Yeomans" to honor his professional achievements.

Chairman SMITH. Thank you, Dr. Yeomans. I was also going to mention the book you wrote that just came out this year called "Near-Earth Objects: Finding Them Before They Find Us." That is a nice subtitle. But I appreciate your writing about this subject, and who knows, maybe anticipating the publicity that subject would have this year as well. Thank you for your testimony.

Dr. A'Hearn.

STATEMENT OF DR. MICHAEL F. A'HEARN, VICE CHAIR, COMMITTEE TO REVIEW NEAR-EARTH OBJECT SURVEYS AND HAZARD MITIGATION STRATEGIES, NATIONAL RESEARCH COUNCIL

Dr. A'HEARN. Mr. Chairman, Members of the Committee, thank you for the invitation to appear today and to discuss a variety of aspects of the near-Earth object hazard. I will talk a little bit less about finding them than Don and Ed had done and more about what to do about it.

Once we complete the George E. Brown survey down to 140 meters, we have taken care of a large fraction of the risk where we can get long advanced warning, and therefore, have plenty of time to mount a mitigation campaign. As was just pointed out regarding the ATLAS survey, that is designed for late discoveries, and as we go to really small ones, late discoveries will be a different kind of issue, because then we don't have time to do mitigation other than an evacuation for 30- to 50-meter city killer—and that is really more than a city. Tunguska was 2,000 square kilometers. You can, in principle, do evacuation, but if you get much larger than that, 75 meters, 100 meters, evacuation is no longer practical, and you need to have a plan in place with tested technologies to try to do mitigation on relatively short notice, because these are likely to be shorter notice than the ones we have been discovering so far where we have been aiming for years of advanced warning and plenty of time to plan how to mitigate.

The mitigation is a key part of the hazard issue, and when the National Research Council issued its report, it suggested programs at a variety of different levels depending on how much insurance you wanted to buy basically. And if you really want to include mitigation as part of that, it is up at the couple of hundred million dollars a year level in order to include mitigation.

Now, it is interesting that most of what we—much of what we know about mitigation so far has come from research programs. They are the ones that provide the physical characteristics. Earth-based remote sensing tells us about the sizes of the different asteroids, tells us about their surface composition, but not necessarily their interior. For a few of them, such as binaries, we can get interior bulk densities. But missions to these objects—we just heard a mention earlier from the Chairman of the sample return mission that will be launched in 2016. That mission will tell us a great deal about the structure of an asteroid—the internal structure and what the materials are, and therefore, what kind of techniques will work efficiently for mitigation.

The Deep Impact Mission, of which I was a principal investigator in 2005, carried out an impact on the nucleus of comet Tempel 1.

It showed first that cometary nuclei are remarkably porous. That makes them harder to push around than, say, a solid iron asteroid. And the rocky asteroids, which are fragmented, are somewhere in between. It demonstrated new techniques for autonomous navigation to lead to an impact. Whether you are doing a kinetic impact or a standoff nuclear explosion, it demonstrated how difficult the attitude control is when you get close to some of these. Milligram pieces of dirt or rock were bouncing our third-of-a-ton spacecraft around by many degrees, causing serious pointing problems. That is an important thing you need to do if you are developing mitigation.

So these research programs are important because they are the only ones that are now providing us information on physical characteristics. Unfortunately, the Discovery Program has been devastated. It was originally conceived as at least one new mission every two years. In the 1990s there were six missions. In the 2000s there were five, the last of which was in 2007. Then, there was a five-year gap until the one that was selected in 2012, namely InSight, the mission to Mars. And with NASA's current plans, the announcement of opportunity for the next one won't be until 2015, which means selection to fly in 2017. So we are down to two per decade instead of the five a decade the decadal survey recommended and which was the basis for the original program. Frequent opportunities to go to space are critical.

Also, just as it is important to partner with the private sector, it is crucial to also partner internationally for mitigation because mitigation can be seen as threatening. And we need to develop real mechanisms. We have talked a lot with potential international partners. We need to be talking to people who aren't our partners such as the Chinese, people who might think something we did in space was a threat rather than trying to help, and that needs to be something that we need to look very carefully at in the near future.

Thank you.

[The prepared statement of Dr. A'Hearn follows:]

TESTIMONY BEFORE THE HOUSE COMMITTEE ON SCIENCE, SPACE, AND TECHNOLOGY

THREATS FROM SPACE: A REVIEW OF PRIVATE AND INTERNATIONAL EFFORTS TO TRACK AND MITIGATE ASTEROIDS AND METEORS

Michael F. A'Hearn
Professor Emeritus
Dept. of Astronomy, University of Maryland
10 April 2013

Mr. Chairman and members of the Committee, thank you for the opportunity to appear today to discuss the potential threats of near-Earth objects (NEOs) in the context of the NRC report on this topic that was issued in 2010. I was the chairman of the mitigation sub-panel for the NRC report, but today I am not representing the NRC, nor NASA, nor the University of Maryland.

The NRC Study: As mandated by Congress in the Consolidated Appropriations Act, 2008, NASA commissioned the NRC to study Surveys for Near-Earth Objects and Hazard Mitigation Strategies. The Steering Committee was chaired by Dr. Irwin Shapiro of Harvard University and the two sub-panels, one for Surveys and Characterization and one for Mitigation Strategies, were chaired by Dr. Faith Vilas, then Director of the MMT Observatory in Arizona, and by myself, respectively. The committee had a wide variety of expertise, ranging over the entire scope of the impact hazard problem. Several public hearings were held, with testimony from numerous experts, some of whom were advocates of specific projects while others were experts in impact prediction and risk communication, and yet others were policy experts.

The committee concluded that the money being expended at that time on NEO surveys was inadequate to meet the congressional mandate of finding 90% of potential impactors larger than 140 m on any reasonable time scale. The committee did not make a specific recommendation on the forward path, but described forward paths for surveys and discovery as a function of how much money Congress wished to appropriate to "buy insurance" against an impact. The amount of money to be appropriated would directly affect the timeline. The committee also recommended initiating a search for potential impactors in the 50-140-m range. The committee noted that there are basically four approaches to mitigation – evacuation for the smallest impactors, slow push-pull techniques, such as the gravity tractor, for moderately sized impactors with long warning times, and then kinetic impactors and standoff nuclear explosions for successively larger impactors and/or shorter warning times. A research program to better understand these mitigation approaches was recommended. Actual mitigation experiments in space were suggested, provided sufficient funding was provided, and overall programs were described for three different levels of funding.

The committee's report, Defending Planet Earth – Near-Earth-Object Surveys and Hazard Mitigation Strategies, was released in 2010. The remainder of this testimony concerns the details of some of these recommendations, both as recommended by the NRC and including my personal perspectives on the issues.

Impactors <140 meters: At the time of the NRC report, results newly published at that time indicated that previous modeling of impacts, by scaling from nuclear explosions of known yield, were incorrect due to the rapid downward motion of an external impactor compared to a nuclear explosion, for which the source can be considered to be at a fixed altitude. These results, which are still neither refuted nor explicitly confirmed, show that substantial damage can be inflicted by objects that are even smaller than 50 meters in diameter. To be specific, the new calculations suggested that the Tunguska event, which in 1908 flattened every tree over roughly 2000 square km in Siberia, was due to a body in the range of 30-50 meters diameter. Based on our knowledge of the size distribution of NEOs, that corresponds to an event that should occur roughly every century or two. For comparison, the best estimate of the Chelyabinsk meteor in February, which caused one building collapse and lots of broken windows with many people injured, is that it had a diameter of 15-20 meters, much smaller than any of the previous estimates of a hazardous size. The size of the Chelyabinsk meteor is better known than most since the trajectory has yielded a reliable velocity and the recovered samples can be used to infer the density of the body. Such an event should occur every several decades. Thus it is clear that objects much smaller than 140 meters are frequent and are capable of significant damage on Earth, although most of these impacts in the past went unnoticed because they occurred over the ocean or over very sparsely inhabited land areas. Detailed modeling of the effects of small impactors, say from Chelyabinsk-size to 140-m diameter, is a gap that should be filled, although most of the computer codes to tackle this problem accurately are under restricted access.

It is widely understood that small objects are much more abundant than large ones in nearly all the populations of the solar system, and specifically among the NEOs. Very roughly, a 14-m NEO is 1000 times more likely than a 140-m NEO. Thus the "next" significant impactor will most likely be closer in size to Chelyabinsk than to 140 meters. It therefore is important to plan for such an event, even if the hazard to life is small.

A key issue for the small impactors is that they are normally so faint prior to impact that we do not know how to detect them very far in advance. Many of them can only be discovered days to weeks before impact. Fortunately, this limitation coincides with the fact that the region of destruction by such an impactor is sufficiently small that evacuation (aka "duck and cover") is a realistic mitigation to minimize loss of life (but not property damage). You will hear about current efforts related to the ATLAS system from Dr. Yeomans and about the private venture to deploy the Sentinel system from Dr. Lu. All other things being equal, space-based systems offer a major advantage in principle, as long as the orbit is sunward from the Earth, such as at Earth's L1 Lagrange point, because it avoids the need for multiple sites on the ground. However, a cost benefit analysis must be undertaken that includes, limiting magnitude, wavelength range of operation, and the data processing approach. The ATLAS system alone is not sufficient for reliable detection because it consists of only two telescope systems, i.e., telescopes at only two sites, but it is designed to be sufficiently low in cost that other countries could realistically deploy similar systems, thus providing 24-hour coverage of both northern and southern hemispheres. The real issue then will be simply implementing the real-time coordination among the systems.

Programs at Various Funding Levels: The NRC report noted that any program dealing with NEO hazards as policy, as opposed to programs dealing with NEOs as scientific targets, should be considered as a form of insurance. The hazard is different from other terrestrial hazards, however, in that the insurance can be used to prevent damage rather than paying for restoration after damage. The question should be thought of, therefore, as a question of how much insurance the nation should by. The committee then described three different scenarios, depending on how much insurance was being bought, with rather arbitrary levels being chosen for the scenarios.

At a level of $10 million per year, the then operating survey programs could continue, as could a modest research program into issues related to the NEO hazard. This level would not meet the congressionally mandated George E. Brown survey to detect 90% of potential impactors larger than 140 meters in diameter.

I note that current spending in NASA's NEO program has increased to roughly $20 million per year, allowing some new initiatives such as the ATLAS program, operations of the PanSTARRS system (currently only one telescope but soon to be two telescopes), and research grants into mitigation related topics. Spending for the Large Synoptic Survey Telescope is not included in these totals – that telescope, if operated in NEO survey mode only, could meet the 140-meter goal relatively quickly.

At a level of $50 million per year, operation of a telescope such as LSST could be funded for NEO-optimized searches, although this assumes construction funding for astronomical research, *e.g.*, from NSF. Alternatively, an in-flight mitigation mission might be feasible if conducted as a minor part of an international partnership.

At a level of $250 million per year for a decade, the advanced surveys to 140 meters could be completed, either from the ground or from space, and a unilateral mitigation experimental mission would be feasible.

None of the NRC's recommended funding levels addressed the question of impactors smaller than 140 meters. With current technology, late detection appears to be the only feasible approach. Limits for the Sentinel system are not readily available to me, nor are the actual limits of the ATLAS system so I cannot comment on their relative contributions. The NEO program office at JPL has funded an independent study to assess the capability of the ATLAS system.

One also needs to remember that, once the George E. Brown survey to 140 meters is complete (90%), the remaining unidentified impactors include both the smaller impactors and the long-period comets. Although the long-period comets very rarely impact Earth, cumulatively they are likely to lead to as many or more deaths as the much more frequent small events. They have been ignored up to this point because they have been such a small fraction of the total threat, but that situation will change dramatically. One has to decide whether to deal with the small, frequent events or with the rare, large events, or both, analogous to deciding whether to deal with frequent auto accidents or infrequent large airliner or ship accidents or both.

International Cooperation and Collaboration: International collaboration is very important in the entire effort to deal with the impact hazard, from discovery, through impact prediction, to mitigation. Unfortunately, despite considerable discussion at the individual scientist level and considerable discussion at the governmental level up to the United Nations, the U. S. is the only nation with a funded, active and effective survey/discovery program. Canada has just launched (February 2013) and Germany will soon launch a small satellite designed to discover sub-populations of NEOs, but the U.S. is still the predominant nation in funding an active program for tracking NEOs, both through the JPL NEO Program Office and through funding the entire operation of the Minor Planet Center that is nominally sponsored by the International Astronomical Union.

It should be pointed out that the only terrestrial impactor ever predicted in advance was 2008 TC_3, an impactor much smaller (roughly 4 meters) than the Chelyabinsk meteor. This was discovered less than one day before impact, by R. Kowalski at the Catalina survey, based in Arizona. The impact was predicted only because the Catalina survey included a (NASA-funded) telescope in Australia in addition to the telescopes in Arizona, which allowed very rapid follow up data, and it was the combination of data from both telescopes that allowed the rapid prediction of the impact, including a prediction of the time and location of impact, both of which were extremely accurate. Thus an internationally distributed, and closely interactive, network of telescopes is critical for predicting small impactors. Fortunately, 2008 TC_3 was so small that it caused no damage on the Sahara Desert in northern Sudan where it entered Earth's atmosphere, although small pieces were subsequently recovered days later.

The area in which international collaboration is even more important is mitigation, due largely to the fact that incorrectly changing the orbit of a potential impactor could merely move the impact site from one country to another, with obvious international implications. Even the Chelyabinsk meteor was claimed by a fringe politician in Russia to be an American weapons test, but fortunately the Russian Academy of Sciences was in the forefront of public announcements, clearly declaring that this was a natural meteor. Unfortunately, there has been even less international discussion on this topic than on the survey/discovery/prediction topic, although there have been discussions within the UN's Action Team 14 of COPUOS. This is an area in which international collaboration, not just discussion, must be established before action is needed.

Contributions of Basic Research to Detection, Characterization, and Mitigation: There is considerable overlap between basic scientific research on comets and asteroids, *i.e.*, on the bodies that include NEOs, and policy-based work on the issues of hazard prediction and mitigation. However, the focus is very different between the two areas and consequently there are significant activities that are not included in one focus or the other. It is for this reason that NEO hazard activities require a separately identified source of funding, associated with national policy, that is not taken out of the scientific programs.

The research activities related to surveys and discovering bodies are aimed at finding statistically significant samples to enable interpretation, and these were the precursors of the specific hazard surveys, which are aimed at discovering as close to all of the objects as is practical (widely being taken to be 90% of the estimated total population). The research surveys, coupled with the work of dynamical researchers studying the orbits of the bodies, are what led to the recognition of the scale of the hazard and many of the individuals involved in those surveys are also involved in the hazard-driven surveys.

Research activities are also directly related to mitigation, but clearly distinct from actual mitigation planning. One of the key issues in mitigation, and for that matter even in predicting the scale of the damage from an impact, is to understand the physical properties of the impactors. Research programs using remote sensing have shown unambiguously that there is a wide variety of physical characteristics among the NEOs, ranging from likely coherent bodies that are the source of iron meteorites through really porous cometary nuclei that are likely to have been the source of the dinosaur-killer K-T impact 65 million years ago. Remote sensing can study a large number of objects and they are sensitive primarily to surface properties of the objects, to their size, and in some cases to a crude measure of their shape and their density.

Important, detailed characteristics of the NEOs can only be learned from *in situ* studies and PI-led, competitively selected missions, under NASA's Discovery and New Frontiers programs, provide the key mechanism to carry out these studies. Such missions can only be used to study a very few targets for budgetary reasons. A team led by Mike Belton and myself proposed the Deep Impact mission to the Discovery Program many years ago purely as a scientific mission, with only two sentences in the proposal about the possible peripheral benefits for NEO hazard mitigation. What the mission did for hazard mitigation was to demonstrate active targeting to impact on a small body, the nucleus of comet 9P/Tempel 1 (a technique needed for our science but also a technique needed for mitigation) and it also demonstrated the very porous nature of cometary nuclei (probably 10% of NEOs are inactive cometary nuclei). The observations of the ejecta were used both to determine the bulk density (much empty space inside!) and to estimate the momentum transfer efficiency of the impact as relatively low (roughly 2), a critical parameter for altering an NEO's orbit with a kinetic impactor. The mission also showed the challenges of attitude control in the last minute of approach to a cometary nucleus. These results have been presented to various groups directly concerned about mitigation, such as the Defense Threat Reduction Agency. The results of the subsequent flyby of comet Hartley 2 as part of the EPOXI mission showed the diversity among cometary nuclei and the heterogeneity from place to place on a single nucleus, both of which must be taken into account in mitigation.

The OSIRIS-REx mission, scheduled for launch in 2016, is a very different mission to a different type of NEO, the asteroid 1999 RQ_{36}. This mission will return a sample of the asteroid to Earth for detailed analysis, but while at the asteroid it will also produce, for example, a detailed map of the gravity. In addition to the material properties learned from the returned sample, gravitational mapping can be used to understand the internal structure of the asteroid, critical information for understanding how to mitigate by changing the orbit, whether by kinetic impact, or nuclear explosion, or even with a gravity tractor, which depends less on the physical structure but does depend on the bulk density and the shape.

These competitively chosen "research" missions are not sufficient to completely address mitigation, but they provide most of the necessary information on the range of physical properties one might encounter. Unfortunately, the NASA budget for planetary exploration has been such that NASA's Discovery program (competitively selected, PI-led missions with a cost cap of $425M in the latest round), have been devastated compared to even a decade ago. The NRC's recent decadal survey of planetary science recommended that NASA's priorities should be first to maintain a cadre of good researchers, and then to maintain a regular cadence, averaging a new start every two years, for the smallest missions (the Discovery Program), then the New Frontiers program (similar to the Discovery Program but for missions twice as expensive), and finally flagship missions (center directed missions that have lately cost more than $2 billion). Although not every mission in Discovery and New Frontiers is relevant to hazard mitigation (the most recent selection in the Discovery program is a mission to Mars), restoring Discovery to the originally intended cadence of research missions would significantly help with the mitigation effort by ensuring the existence of other missions to comets and asteroids to provide information necessary for mitigation.

Ultimately, however, specific mitigation missions must be considered as discussed above under program levels. They should be funded over and above the research program and they could be either separately funded add-ons to scientific missions or stand-alone missions, or international collaborations, with the international collaboration a high priority. Note that once the range of physical properties is understood, it is still very difficult to determine the physical properties of an actually threatening NEO without sending a mission to it, a possibility with very early discoveries but not with late discoveries.

What Should be Done in the Event of an Identified NEO Threat? After an NEO threat is identified, the initial steps are well defined. NASA is the lead agency for identifying threats and they have a reporting path through the U.S. government that covers all relevant federal agencies and the POTUS. Reporting to other countries is also urgent and should be done through the U.N. in order to reach all governments. In addition, there should be direct communication with countries and international agencies that have relevant capabilities for mitigation. Immediately following the alert, it is crucial to share all available data publicly. This is routine for the positional observations of the NEO and for the resultant orbital computations through the Minor Planet Center and through JPL's NEO Program Office. Beyond this, however, it is crucial to share all available information on the physical characteristics of the NEO from whatever source and on the details of the impact prediction. In the case of 2008 TC_3, which presented no hazard, this information was communicated through the channels normally used worldwide by astronomers and information was made readily available to news media.

The next steps depend critically on the nature of the threat – how big the impact will be, how far in the future it will occur, and where it will occur. An all-out effort to determine the characteristics of the particular impactor is crucial – remote sensing being needed in any case and, if time permits, a mission to characterize the NEO should be initiated in order to optimize the mitigation. Short warning times, however, may preclude an advance characterization mission and in that case the range of expected properties must be used to design a fail-safe mitigation. Action paths are, to my limited knowledge, not yet in place domestically. For a small impactor, a plausible route is through FEMA. For a larger impactor, however, either the military or NASA might be the one to take charge. For truly large impactors, the lead country and agency should be coordinated among those countries that have the capability to execute any mitigation. This decision/action tree should be fleshed out and made publicly available long before any specific threat is identified.

Biographical Information

Michael F. A'Hearn

At the University of Maryland, Michael F. A'Hearn is a Professor Emeritus and Research Professor of Astronomy, having retired as a Distinguished University Professor of Astronomy in 2011. His research has emphasized the observational and experimental study of comets and asteroids and what they can teach us about the origin of our own planetary system. He has also taught a wide variety of courses in astronomy.

Dr. A'Hearn has played leading roles in observational studies of comes and asteroids including many of the early measurements of asteroid diameters via occultations and the most extensive survey of the composition of comets from ground based observations. He has worked extensively with telescopes in orbit dating from the earliest days of ultraviolet astronomy. Using observations at all wavelengths, he participated in the first discovery of several molecules in comets and he has used the chemical composition to study the origin of comets. More recently he was the principal investigator for NASA's Deep Impact mission, the only mission ever to target a high velocity impact at any of the solar system's small bodies, excavating a crater in the nucleus of comet Tempel 1 on 4 July 2005 that was tens of meters deep and some 40-50 meters in diameter. He was subsequently the Principal Investigator for NASA's EPOXI mission, which studied extrasolar planets and flew past comet Hartley 2 in November 2010. He is a member of two instrument teams on ESA's Rosetta mission and is the PI for the Small Bodies Node of NASA's Planetary Data System.

Dr. A'Hearn holds a BS in physics from Boston College (1961) and a PhD in astronomy from the University of Wisconsin (1966). He has authored more than 200 papers in refereed scientific journals in addition to a variety of other articles. NASA has twice awarded him the Medal for Exceptional Scientific Achievement (2006 and 2012) and in 2008 he received from the American Astronomical Society's Division for Planetary Science the Kuiper Prize for a distinguished career in planetary science. He has taken leadership roles on several panels for the National Research Council of the National Academies, including chairing the Mitigation Panel of the Committee to Review Near-Earth Object Surveys and Hazard Mitigation Strategies.

Chairman SMITH. Thank you, Dr. A'Hearn. I will recognize myself for questions.

Dr. Lu, given the fact that we do have budget constraints and that funding is limited, what is the most—single most important thing we could do in this, say, next three- to five-year time period to detect these threatening asteroids?

Dr. LU. Well, if you are going to ask what is going to find the most number of these asteroids that—of anything that is currently planned, I think it is Sentinel pretty—by a pretty good margin. And if you were to ask, you know, to get to what Congressman Johnson mentioned, which is, you know, we are a private organization——

Dr. LU [continuing]. There are opportunities to accelerate our development. You know, we could, in principle, deepen our relationship with our currently existing public-private partnership if we wanted to accelerate that. We understand again that the technology that we are developing—the core technologies are useful for lots of other things that the Federal Government finds important, and so, you know, one of the possibilities is to accelerate the technologydevelopment.

Another possibility is if this data is worthwhile to NASA, if it is important to NASA, perhaps we could work out something where this data is purchased from us, and that way NASA only pays for it if the data is good and they could work with us to make sure that the—as they already are, that the quality of the data is what they need.

Chairman SMITH. Okay. Thank you, Dr. Lu.

Dr. Yeomans, I am sure there is an answer to this that I should know, but we have always been told that in the case of near-Earth objects, the only alternative is to move them out of their orbit or deflect their trajectory so that there is no direct impact, and it doesn't do any good to explode those objects because then we just get a shower of near-Earth objects, many more but they are smaller. Is it—would it be possible to explode an incoming asteroid with such force that the pieces would be so small that they would burn up coming into the Earth's atmosphere? So is that a realistic alternative or not?

Dr. YEOMANS. Yes, it is actually. There has been some work done by Dave Dearborn at Lawrence Livermore Laboratories using computer simulations. If you insert an explosive charge and detonate it, you often get the fragments going off at such velocities and directions that what little does hit the Earth does so with very little damage.

Chairman SMITH. Why is so much, therefore—so much time, so much effort, so much focus on moving it out of its current trajectory? Why not more focus on what you just described?

Dr. YEOMANS. Well, it is actually considerably easier to run into it and slow it down a tiny little bit than to land on it and plant an explosive device and——

Chairman SMITH. Okay. Another practical answer as well.

Dr. YEOMANS. It is technologically easier.

Chairman SMITH. Okay. Dr. A'Hearn, you mentioned about the time with—that would—we would have or not have if we detected an incoming object and had to deflect it. What would be the average time that we would have, say, of a city killer-sized asteroid? I guess it depends on whether you are using ground-based telescopes or space-based telescopes, but say in the next three to five years, how much time would we have if we developed the Sentinel program and were able to detect these objects?

Dr. A'HEARN. I will defer questions on the sensitivity of Sentinel to Ed Lu——

Dr. A'HEARN [continuing]. But in general, it is important to remember we have only ever detected one incoming object before it hit.

Chairman SMITH. We have a long way to go——

Dr. A'HEARN. That was less than a day out.

Chairman SMITH. Okay. Dr. Lu——

Dr. A'HEARN. And that was very small.

Chairman SMITH. Do you have any ideas on how much——

Dr. LU. The goal for Sentinel is to find things decades before they hit so that you can deflect them rather than evacuate.

Chairman SMITH. We have plenty of time. At our first hearing, Dr. Holdren made the point, I think, that only two percent of the Earth's surface consists of urban areas and so that further diminishes the possibility of a city sustaining a direct hit. I am not sure that is much consolation to those who live in rural areas by the way, but at least it was interesting as far as the amount of damage that might occur.

But thank you all. You have answered my questions.

And the gentlewoman from Texas, the Ranking Member Ms. Johnson, is recognized for hers.

Ms. JOHNSON. Thank you very much.

Dr. Lu, I realize that details over NASA's proposal in its FY 2014 budget request to conduct a mission to an asteroid with humans and other asteroid-related activities are just trickling out. A story over the weekend reported concerns about the asteroid initiative from two sources. One worried that NASA's activities may interfere with the private-sector efforts. Another was critical of the absence of international collaboration. Based on what you have read or know of NASA's plans, are such concerns warranted?

Dr. LU. I don't think so. I believe that—you know, I, as much as anybody, want our human spaceflight program to have a clear, defined, and inspiring goal. However, I don't think—this mission should not be confused of one that is planetary defense. That is a very—it is a different mission——

Ms. JOHNSON. Um-hum.

Dr. LU [continuing]. What that proposed mission is to do.

Ms. JOHNSON. Dr. Yeomans, can you share details on NASA's asteroid detection effort or efforts that are scheduled to benefit from the increase for the coming fiscal year?

Dr. YEOMANS. Yes, it is my understanding that the asteroid retrieval mission is primarily a technology test of the solar electric propulsion system. It is also a rendezvous with a small asteroid with an attempt to bring it back into a lunar orbit. It has components for NASA's human exploration program. And, of course, the most challenging first part of this whole mission idea is to find a suitable target. So the plus-up that you mentioned in the budget will certainly provide a commensurate increase in the number of objects that are discovered and could be utilized for space resources, scientific investigations, planetary defense, as well as a target for this mission.

Ms. JOHNSON. Dr. A'Hearn, did you have any comment?

Dr. A'HEARN. No, I have no further comments.

Ms. JOHNSON. Thank you. Now, in the first round of hearings that we had, there was mention of an orbiting telescope, which we don't have access to. Could either of you comment on the value of having an orbiting telescope?

Dr. LU. Well, the Sentinel is an orbiting telescope——

Ms. JOHNSON. It is.

Dr. LU [continuing]. But it does not orbit the Earth. It orbits the sun. But it is a space telescope.

Dr. YEOMANS. There is also——

Ms. JOHNSON. Yes.

Dr. YEOMANS. There is also a concept where you have a spacecraft a million miles sunward of the Earth also orbiting the sun, but it is closer to Earth and could be looking out toward near-Earth asteroids as well.

Ms. JOHNSON. So in view of the seemingly increased interest for activity of the asteroids, how do you see an investment in an orbiting telescope that would orbit the sun or in a place it is not orbiting now? Do you see any value?

Dr. YEOMANS. Oh, yes. Yes, as Ed mentioned, the benefit of having a telescope in space is several-fold. First of all, you can use an infrared detection system, and these objects are much brighter in the infrared and much easier to find than in the optical region. You don't have problems with weather or day and night. You can observe these objects from a viewpoint that the Earth cannot, so you sometimes get an advanced warning in that respect. So it is a far more efficient system from space.

Ms. JOHNSON. Any other comment?

Okay. Thank you, Mr. Chairman.

Chairman SMITH. Thank you, Ms. Johnson.

The gentleman from Texas, Mr. Hall, the Chairman Emeritus is recognized for his questions.

Mr. HALL. Mr. Chairman, thank you.

I probably ought to just write a book on this, my questions, because I have so many. And this is your second hearing, is that right?

Mr. HALL [continuing]. And I admire you for it. You are almost seeking something that is impossible from the numbers even than that I have heard here.

Olin "Tiger" Teague, whose picture is right over there, ought to be known as the father of NASA. And you might even become the father of characterizing Earth objects and how close they are. But I think you are going to have to have some overseas hearings. We have had this second one and four more just like this probably won't yield any definite answers. But it is a very interesting matter, a very interesting item.

But how could we ask—and maybe, Dr. Yeomans, on to the near objects program, like in the 2013 worldwide attention in the city of Chelyabinsk in Russia—injured a lot of people but didn't kill anyone is what I understand. And the others that I remember and that we have heard about, a lot of injuries, but what—they didn't know it was coming and didn't know what we—when we had that hearing—and I think I testified to this before—we found out in that hearing that one had passed Earth and just missed us by 15 minutes. And that could be a jillion miles away, but that is where they put it. It had missed Earth by 15 minutes, and nobody even knew it was coming until it had come and gone.

So—and we made every effort to get in touch with nations like Japan, Spain, Italy, England, France to send somebody over here to testify with us because it has to be a world for us if we are going to really do anything about it. We can't pave the way like we spent 34 billion on global warming. My President spent 34 million on that and haven't done anything on it.

But it looks like we are going to have to have world input to ever be anywhere near efficient on making the determination that all the people in the world that I want. But we couldn't get any interest at all. And I think this Chairman of this Committee that that would be a very good thing if you could have some hearings, maybe in England, be at the places and get their interest up because it is going to take their working with us to make anything happen.

I guess the only question I would have is, Dr. Yeomans, whether you know of any private organizations that are involved in near-Earth object detection like Boeing or Lockheed or McDonnell Douglas or Texas Instruments? You know, I would like to get them into it but they can't do it themselves. So that is just something to think about. Do you have any suggestions on the private organizations and how they might work into it?

Dr. YEOMANS. Well, Ed——

Mr. HALL. They said at one time a laser could affect them just a little bit but it didn't say how much.

Dr. YEOMANS. That is true. If you had a laser nearby, you could ablate the front side and introduce a thrust in the opposite direction. But in terms of the international cooperation, I couldn't agree more. In fact, the European Space Agency has been getting more and more interest in this near-Earth asteroid discussion. Recently, they are actually funding the so-called NEOShield program to look at various mitigation options, including kinetic impactors.

There is an activity within U.N., COPUOS, the Committee on the Peaceful Uses of Outer Space, and NASA is involved with that working group to try and define an international warning system, with the response protocols that would be required in the event of an incoming object. Who would be in charge? Who——

Mr. HALL. Dr. Lu suggested NASA. The reason I thought about "Tiger" Teague, Olin Teague, and all the work he did in even getting it off the ground and supporting it with funds that we don't have today. And we can't go to Mars until people can go to the grocery store, so I don't know how we are going to talk about protecting the world if we don't have world support. And it would be a great thing for this Chairman if the government doesn't have the money to send them, he has personal wealth if he could maybe take four or five of us over there. And I think he is going to tell me my time is over. I yield back what time I do have.

Chairman SMITH. Thank you, Mr. Hall.

The gentlewoman from Connecticut, Ms. Esty, is recognized for her questions.

Ms. ESTY. Thank you very much. I wanted to follow up a little bit on—we have had discussion previously about this international issue which, Doctor, you had mentioned. Can you explain to us what is currently done in terms of data sharing? And perhaps, Dr. Lu, if you could discuss if you have even contemplated now for your project, it not being a governmental project, being private nonprofit, what you would contemplate being that data-sharing aspect for your organization?

Dr. LU. Yeah, I—you know, our intention as a public nonprofit is to put the data out there so as many scientists can see the data and use the data and warn people if there is things in the data that show that something is going to hit the Earth. So that is our plan.

But actually, if I could add one other thing. Don't get the impression that finding asteroids—while it is a lot of money—is something that requires enormous amounts of money. For instance, I mean our telescope, which will find and track a great majority of these asteroids, is less than the cost of—there is a road-widening project in the San Francisco Bay area in the town of Burlingame that is more expensive than our telescope. And that is why we went about this as a private fundraising effort. We are less expensive than a museum. There is a wing of an art museum in San Francisco that cost more than our project. And that is privately raised money. It is not enormous. I mean, it is a lot for individuals, but it can be done.

Ms. ESTY. If I can follow up, actually. That was very helpful, because I did want to ask a little bit more. What are the specifics? What are your plans if you—obviously, we know that in the past NASA has encountered cost overruns for a variety of reasons. What are your plans as an organization if you discover, say, in the development or in the research phase that something you anticipated will work does not quite work the way you expect it? Would you go back to funders? What does that do? How—and also, frankly, how close are you to raising the $450 million that you have budgeted? When will you start? Will you do it in tranches if you don't have it ready? What are your plans for ensuring that? Because we are hearing from everyone if you don't have the money set at the outset, you end up embedding cost overruns because it just takes longer.

Dr. LU. We are using existing technology to the extent we can, and we actually have a firm fixed price contract, which is—so in other words, the risk is borne by our contractor Ball Aerospace. And yes, we are raising the money in tranches. This year, our goal—fundraising goal is $20 million. And we are well on our way towards that for this year.

Ms. ESTY. So how much of the total do you have—would that have you at?

Dr. LU. Well, this really is our first full year since we have begun our fundraising. We announced on June 28 of last year. Our needs were quite small last year, in the single-digit millions. This year, they are accelerating, and next year they will accelerate even more, so our peak spending rate will be in the range of $100 million for a year or so, and then it will taper back down. But we can finance this over a much longer period.

Ms. ESTY. And just a question for all of you somewhat. If we have congressional mandates that, say, previously would have been directed to NASA as a governmental organization and Congress says we need to see this data because we need to make decisions, for the doctors who are not in the private entity, how would you contemplate we should—would structure that? And how would that operate?

Dr. YEOMANS. Well, I think it is important to point out that NASA does have a Space Act Agreement with the Sentinel group for providing navigation and tracking of their spacecraft. Once their data are taken, it would come through the NASA channels. It will go to the Minor Planet Center in Cambridge, Massachusetts, then it would come to our program office at JPL, and then we would interact with our Italian colleagues and we would post our results for the world. So it is quite a transparent data-sharing process even though it is privately funded, for the most part.

Dr. A'HEARN. Yes, I was just going to comment that in my experience the data on finding and tracking near-Earth objects and on predicting the orbits of them all becomes very public very quickly. There has never been a problem getting the data. The only problems are what to do with it.

Ms. ESTY. Thank you all very much.

Chairman SMITH. Thank you, Ms. Esty.

The gentleman from California, Mr. Rohrabacher, is recognized for his questions.

Mr. ROHRABACHER. Thank you very much. I will try to be as fast as I can here.

First of all, I would like to note——

Chairman SMITH. And Mr. Rohrabacher, if you will suspend for a minute, I want to let the Members know that after this series of questions, we are going to recess for about 45 minutes so we can go conduct three votes, and then we will resume the markup after the votes.

And the gentleman continues to be recognized.

Mr. ROHRABACHER. Great. Thank you, Mr. Chairman. And number one, first of all, Mr. Chairman, I would like to agree with Chairman Hall and his recommendation that we work with you and Members of both sides of the aisle to try to find international cooperation on an effort that deserves to be not just the responsibility of the American taxpayers but people of the Earth united against this common threat.

Let me note there are other groups like the Planetary Society, headed up by Bill Nye, who are very involved with this issue. And I have a statement that I would like—of Mr. Nye that I would like to put in the record at this point.

Chairman SMITH. Without objection, so ordered.

[The information may be found in Appendix II.]

Mr. ROHRABACHER. Thank you very much.

Next, I would mention there are two recently formed companies that have as their goal mining asteroids: the Planetary Resources, Deep Space Industries. Both of these companies have impressive teams, and I would hope that at some future date we might be able to bring them to testify about their activities and the expertise that they are developing.

Dr. Lu, I found your testimony to be very interesting. We have to assume that either road construction in San Francisco is incredibly expensive or that we have in some way brought down the cost of your efforts—space efforts. I find—and it was your testimony that B612 does not in any way receive any taxpayer funding?

Dr. LU. That is correct.

Mr. ROHRABACHER. Congratulations, Dr. Lu. I want to say that for the record, congratulations. And I understand that the Sentinel mission under the Foundation actually has been operating with fixed-term prices that you are dealing with your—with the companies that you have to deal business with. Is that correct?

Dr. LU. That is correct.

Mr. ROHRABACHER. So you have a fixed-price term. We have been told over—again and again, Mr. Chairman, that we can't have these fixed-price contracts. For example, with our polar weather satellites, oh, you can't have a fixed-price contract. Perhaps this private sector group here that doesn't receive any of our government money is showing us how we can keep some of the costs down.

And let me just suggest that we need to get more private money, more international cooperation. This is a serious threat to the—not only to the well-being but even, perhaps, to the survival of humankind on this planet, and it deserves us to work together and to do so in a cost-effective way. And we can't do anything nowadays unless it is in a cost-effective way.

I would like to thank you, Mr. Chairman, for holding this hearing, and I just will leave it at that. And I appreciate your efforts and am totally supportive.

Chairman SMITH. Okay. Thank you, Mr. Rohrabacher. And I know this subject has been of long-time interest to you as well.

As I say, we are going to recess for about 45 minutes, and then I hope Members who still have questions will return. And if you all can possibly stay, that would be great. I understand one witness may have to leave, and if that is the case, we understand that as well. So thank you all, and we will return and we will recess until about 45 minutes from now.

[Recess.]

Mr. PALAZZO. [Presiding] I want to thank the witnesses for staying behind for this important Committee hearing.

And at this time, I am going to recognize Ms. Bonamici for five minutes.

Ms. BONAMICI. Thank you very much, Mr. Chairman.

Thank you so much for your testimony and thank you for staying. Sorry we had to leave to vote.

I wanted to talk a little bit about how we respond. And Dr. A'Hearn, in your prepared statement, you indicate the Academies' 2010 report provided options geared to how much money Congress wished to appropriate to buy insurance against an impact, and you described evacuation for small impactors is one approach to mitigation and noted the panel's recommendation that a research program be instituted to better understand mitigation approaches.

I represent a district in Oregon that contains coastline, and my constituents on the coast are frequently talking about being prepared, emergency preparedness for tsunamis and earthquakes, and so these are certainly analogous situations.

In our prior hearing, there was a discussion about evacuations in response to a meteor incident. So what would be the nature of the recommended research as it applies to evacuations? I know that when we are talking over in the Oregon coast now they don't have a lot of time from the time they find out about a tsunami to get upland. So what do you see as the most cost-effective insurance, and can you talk a little bit about preparing for a meteor impact, please?

Dr. A'HEARN. I think the most important issue is that we don't have a really solid theory of how big a tsunami you will get from a given size impact. There are simulations that disagree by huge factors on how big a tsunami you will get at various places. So on that specific issue, I think that is the key thing that needs to be done. It depends on the size of the impact or, of course, depends on the velocity it comes in, the speed, and it depends on the density. You know, is it really solid or is it mostly porous? But for any given case even, there are disagreements in the theoretical literature on what the effect will be.

So that is the biggest issue. Once you know how big the tsunami will be, then you will get a better feeling for how far you have to evacuate to get to high ground. And I am not familiar with how much time is needed in any specific area.

Ms. BONAMICI. Sure. That is dependent, I think, on the geography.

And to all the panel members, how should the policy—how should we approach the policy and legal issues in addressing warning the public? My constituents at home are worried about finding a job and about too many kids in the classroom, so—and on the coast, they are worried about a tsunami and they went through, you know, after the earthquake in Japan, some emergency preparedness, but there is still a lot to do. So what is the best way for us as policymakers to approach this warning and preparedness, and how should we handle that on national and international levels? What is your advice?

Dr. YEOMANS. If I could respond. There is an ongoing effort within the United Nations' Committee on the Peaceful Uses of Outer Space to address these issues, and one of the key issues, as you noted, is how do we best warn the public, give them the facts without scaring them? So on the international level within this Committee, these discussions are ongoing. And that is one of the issues that is front and center. We don't have a process in place. I mean, we are scientists so we can say we are going to impact probably at such and such a time, but that is not necessarily the most effective communication with the public. So we have to bring in folks who are more experienced in communicating risks, not just scientists. I would suggest that perhaps once these discussions are completed in, hopefully, another year, then effective communications would come out of that.

Ms. BONAMICI. Dr. Lu, do you have any input on——

Dr. LU. Yeah, my opinion is that we should not find out what the impact of a large asteroid is in the ocean and—because we have the technology to prevent that.

Ms. BONAMICI. Um-hum.

Dr. LU. And we should go out there and find these asteroids, find out if any of them are going to hit us, and the deflect it. And I think we can do that.

Ms. BONAMICI. Thank you. And my time is about to expire. Thank you very much. Thank you, Mr. Chair.

Mr. PALAZZO. I now recognize Mr. Posey for five minutes.

Mr. POSEY. Thank you, Mr. Chairman. Somebody mentioned climate change study a little while ago. You know, asteroids took care of that at one time, and if it happens again, we will not have global warming. They can fix that forever.

Out of curiosity for the three of you, the Administration is excited about privatizing space to the greatest extent possible. What do you think would be an appropriate number for an X prize type of arrangement for identifying and destroying an asteroid, just off the top of your head, all three of you, starting with Dr. Lu?

Dr. LU. Well, if you ask the question—I mean what would it take to find these asteroids first for the first part of the X prize. It is really a two-step process.

Mr. POSEY. Right.

Dr. LU. I would lay a number out that would be equivalent to whatever—you know, some fraction of what NASA would have spent if they did it themselves. And that number is probably in the range, according to the NRC report, $800 million to a billion. So pick some fraction of that. That is why we think we can do it for $450 million, and that is what our contract specifies. But if you put the prize somewhere around there, then NASA is guaranteed to save money if it succeeds.

Mr. POSEY. Yeah, and if it doesn't, the money is never spent.

Dr. LU. Exactly.

Mr. POSEY. Okay. How about—that is to find one. Does that include destroying it?

Dr. LU. No, but I think if you—once you find them, remember that you will now know if there is something that is going to hit that is a definite threat in the next century. And now you have got time to do it right. And also I think money is also no object if something is really barreling down on the Earth and you know the time, date, and place that thing is going to hit. I think we can come together and solve that issue.

Mr. POSEY. Okay. Thank you. Dr. Yeomans?

Dr. YEOMANS. I would add that NASA already has 15 years of experience in this area of identifying objects. They have three programs underway, ground-based optical detection. I would suggest perhaps a study that could be undertaken to see whether we could leverage those assets to improve what is already there by bringing online new technology and new telescopes along with studies to flesh out what is the most effective way of deflecting an object that is found on an Earth-threatening trajectory.

My comment would be, we should leverage existing activities and facilities.

Mr. POSEY. Okay. Well, it is my understanding the Small Bodies Assessment Group at Lunar Planetary Institute was chartered for the specific purpose of evaluating those types of missions and the priorities of the scientific community for near-Earth objects. How has NASA collaborated or leveraged its information with this group in planning of the Asteroid Capture Mission?

Dr. YEOMANS. I am not intimately involved with the connection between the Small Bodies Assessment Group and this mission that you mentioned. So I am not aware of what has and what has not been communicated between those two.

Mr. POSEY. Okay. Are either of the others familiar with it, Dr. A'Hearn?

Dr. A'HEARN. I know essentially nothing more about this mission than I have read in the newspapers and in Administrator Bolden's release this morning. I am not aware that the Small Bodies Assessment Group has been given any information on it. They may have been, but I am not aware of it, so I am not going to comment further.

Mr. POSEY. That was my feeling and that is why the question. Dr. Lu.

Dr. LU. I also am not aware of the connection between the two.

Mr. POSEY. Okay. You know, all of your written testimony mentioned obviously the asteroid mitigation, and I know we have to identify them before we can divert them or destroy them. We all knew that. But, you know, assuming that the development of a strategy and technology would take a considerable time, you know, obviously perhaps years, what steps do you think we should be taking in the meantime in case our search uncovers a threat, which we all know is not a matter of if but when?

Dr. LU. I think it would be prudent to do a deflection demonstration mission, pick an asteroid that you know is not anywhere near hitting the Earth and show that you can deflect it in a controlled manner so that it doesn't break up into pieces where you don't know where they are going and so on. I think that can be done.

Mr. POSEY. Okay.

Dr. A'HEARN. I would agree that a demonstration deflection mission is an appropriate thing to do, and a deflection mission is ideally suited for the international collaboration that I think is needed in this area, because typically you need to send two spacecraft, one of which does the deflection and the other of which monitors the effectiveness of it. Depending on whether you are doing a gravity tractor or kinetic impactor or—we presumably would not do a nuclear one as a test and the ability to have international collaboration on coordinating two spacecraft is important to get the various countries trusting that we are not trying to divert something to land somewhere else.

Mr. POSEY. Interesting. I hadn't thought about that but I think you are correct. Dr. Yeomans, Mr. Chairman, can he finish?

Dr. YEOMANS. Can I add something? There is an interesting concept pertinent to your point whereby NASA would use the excess launch capability for the InSight spacecraft to Mars, have a co-launch of an impactor much like the Deep Impact mission, and that would go and collide with the asteroid that the Osiris Rex mission has already picked for their target. So the Osiris Rex mission is already resident, and you would have this impactor coming in, and you can measure the deflection. So it is a nice leveraging of an existing launch and an existing rendezvous spacecraft. So that would be one instructing deflection demonstration.

Mr. POSEY. Very good. Thank you. Thank you, Mr. Chairman.

Mr. PALAZZO. You are welcome. I now recognize Mr. Stewart for five minutes.

Mr. STEWART. Thank you, Mr. Chairman. Gentlemen, thanks for being here. It gives me faith in our future knowing that there are people a lot smarter than me who are working on some of these things.

I am not going to ask in real detail. I would like to just kind of encapsulate what I think we have said but bring some clarity to it before with some very quick questions. But before we do, can I just divert for just a second with this, and that is, you know, the old formula $E = MC2$, and you have talked a lot about the mass of these meteorites, potential, you know, objects, but is velocity a consideration, too? In other words, are some of the smaller ones, are they traveling at such a speed that they would have an equally devastating outcome or are most of these objects kind of traveling at about the same speed out there?

Dr. LU. Most of them are—well, they are orbiting the sun, so the typical velocities that they hit is really independent of the size of the asteroid, and that is between 15 and, say, 25 kilometers per second.

Mr. STEWART. Okay.

Dr. LU. So 40,000 miles an hour or so.

Mr. STEWART. So that is—I mean that is a fairly good range. Fifteen to 25 is, what, 40 percent or something like that? But their velocity doesn't really matter. It really is just the size and the weight of the object?

Dr. LU. Well, it is a combination of the destructive power, it is a combination of the speed and the mass. But from the standpoint of deflection, it doesn't much matter.

Mr. STEWART. Okay. Yes, sir, Dr. A'Hearn.

Dr. A'HEARN. I was going to just add to that. Indeed, 15 to 25 kilometers per second is the right ballpark for the asteroids. It is one of the things you have to keep in mind, however, if you deal with the cometary impact hazard. Those come in at more like 30 to 70 kilometers per second. Now, they are very infrequent compared to the asteroids, but one of a given size will be much more damaging because of that high speed of entry compared to the asteroid.

Mr. STEWART. Yeah, okay. And I appreciated your visual that you showed us at the beginning. It kind of gives us a sense of the scope there.

I know there was a recent comet that was discovered in January that was looking like it was going to have a near miss with Mars, and it would have been a devastating event for—had that, you know, impacted the Earth, a dinosaur killing type event. And as I recall, it was two years is what the, you know, estimated impact time would be. Of course, we know it is going to miss it now. If that had been directed toward Earth in two years, is there realistically anything we could have done?

Dr. LU. It would be very difficult.

Mr. STEWART. Probably not, is that true?

Dr. LU. Yeah.

Mr. STEWART. So can you give me an idea? I know you are speculating, but I mean what—how much time do we need? Do we need 10 years. Do we need 20? Do we need eight? I mean, how long do we need before we could actually do something even if we detected an object that was going to impact the Earth?

Dr. LU. I think with 10 years you can do this in a controlled manner with backups and so on. Certainly, with 20 years you could do that. It gets much more difficult the closer in it is, and that is, again, the importance of getting early warning, because the closer it is to you, the more you need to deflect it by to get it to miss.

Mr. STEWART. Yeah

Dr. LU. So it gets much, much harder the earlier—the less warning you have.

Mr. STEWART. Let's put that aside, that consideration of the energy to deflect it. In two years from now, could we—are we technologically capable of launching something that could intercept it? Dr. A'Hearn, you seem to be shaking your head "no."

Dr. A'HEARN. No. If we had spacecraft plans on the books already, that would take a year—I mean a typical small mission like a Discovery class mission takes four years from approval to start to launch. Okay. Now, a really accelerated military program would be faster than that but that is a couple of years still.

Mr. STEWART. Yeah.

Dr. A'HEARN. And you would have to have something ready to launch, basically, if you wanted to do it on very short notice. Ten years, 20 years, then you have got time to plan it. Five years or less, it is really hard unless you have thought the problem through and design things, maybe have components built, maybe have a full system but——

Mr. STEWART. Because what we need, we have nothing like this right now. We are not taking an existing weapons system or existing vehicle and modifying it. We are really starting from scratch to do this, true?

Dr. A'HEARN. Well, you would try to use it from existing components. I mean you could—you would—if you were going to do a kinetic impact, you might scale up what was done for Deep Impact to larger launch vehicle, larger impactor, and things like that. So it is not quite starting from scratch, but it is starting from a pretty low point.

Mr. STEWART. Yeah. Okay. And then last question—well, I tell you what, I am out of time. I would love to talk with you further, but I appreciate you—again you being here. Thank you.

Mr. Chairman, I yield back.

Mr. PALAZZO. I want to thank the witnesses for their valuable testimony and the Members for their questions. The Members of the Committee may have additional questions for you, and we will ask you to respond to those in writing. The record will remain open for two weeks for additional comments and written questions from Members.

The witnesses are excused, and this hearing is adjourned.

[Whereupon, at 4:13 p.m., the Committee was adjourned.]

Appendix I

ANSWERS TO POST-HEARING QUESTIONS

ANSWERS TO POST-HEARING QUESTIONS

Responses by Dr. Ed Lu

Response by Edward Lu, CEO B612 Foundation to questions following the April 10, 2013 House Committee on Science, Space and Technology hearing on "Threats from Space".

Response to questions from Rep. Steven Palazzo:

1) Do we have the tools and technology necessary to detect Near Earth Objects (NEOs)? Once we identify an object, what are our means of tracking it?

Response: Currently deployed asteroid detection and tracking systems are only able to find at most about 1000 Near Earth Objects (NEOs) per year. So they are insufficient to find and track the 1 million NEOs large enough to destroy a city. The B612 Sentinel Space Telescope, will be able to discover and track over 100 thousand NEOs each year, and so will be about 100 times more effective than all current systems combined. By repeatedly observing these asteroids, we can accurately measure their orbits. The orbits, once measured, are stable on roughly a timescale of 100 years, so asteroids in general do not need to be continually tracked once their orbits are determined.

2) What categories of NEOs do we currently track, and which of these present potential cause for concern?

Response: The NASA Spaceguard survey has successfully found over 90 percent of asteroids 1km or larger, which is large enough to wipe out human civilization. The Spaceguard survey is progressively more incomplete for smaller asteroids. Asteroids of size 40 meters are large enough to destroy a city, and the Spaceguard survey has only found about 0.5% of these. In other words, the vast majority of asteroids capable of doing great harm (destroying a city or more) are not currently tracked.

3) Both Dr. Holdren and Administrator Bolden testified to our committee in March that we have a long way to go to accomplish the goals established by Congress in the NASA Authorization Act of 2005 of detecting 90 percent of the NEOs with a diameter of 140 meters or greater by 2020. What are the most important steps that should be taken in the next five years to accomplish these goals?

Response: As described in the National Academies report "Defending Planet Earth", the most effective thing we can do to finding these asteroids is to deploy an infrared space telescope into a Venus-similar orbit around the Sun.

4) Can you describe the potential range of damage caused by impacts from NEOs?

Response: At the small size range, an 18 meter rocky asteroid like we saw over Chelyabinsk on February 15, 2013 is capable of causing structural damage to buildings due to the shock wave produced in the atmosphere. A 40 meter asteroid like we saw in Tunguska in 1908 would cause a multiple megaton explosion, and is capable of completely obliterating a city. A 140 meter or larger asteroid would have an explosive energy of more than 100 megatons, and is capable of destroying an area equivalent to a small state. Depending on where these asteroids hit, the consequences could range from minor up to global economic collapse. A 1km asteroid would likely end human civilization, no matter where on Earth it struck.

5) What are the areas in which there is a lack of knowledge or understanding of near Earth objects? What barriers does the private sector face in gaining the knowledge necessary to quantify and mitigate the risk of NEO impacts?

Response: There is much scientific research still to be done to understand NEOs. But from the standpoint of protecting the Earth from asteroid impacts, the first and foremost thing that must be done is to find and track these asteroids. All other questions are secondary since we cannot defend ourselves from an asteroid we have not found yet. The B612 Sentinel Space Telescope will be able to find these objects. We are a private philanthropic organization and our principal barrier is the speed at which we can raise donations.

6) From where you operate, how would you describe the level of coordination between governments and outside organizations? What improvements need to be made?

Response: The B612 Foundation has an agreement with NASA in which NASA provides no money, but agrees to allow use of its Deep Space Network to communicate with Sentinel. In exchange, the B612 Foundation will make this data generally available to the public.

7) What can the U.S. government do to encourage the advancement of private sector technologies that detect and track NEOs?

Response: The B612 Foundation would be interested in exploring ways in which it could work with the Federal Government to accelerate the deployment of Sentinel. There is an opportunity to show US leadership in protecting not only US territory but the entire world, and to do it in an innovative fashion that furthers commercial and nongovernmental space exploration. Not only would this benefit the world, but the fundamental technology of Sentinel which allows it to accurately track small dark objects in space would be useful in specific ways to several agencies of the U.S. government.

8) Should the cleanup of space debris primarily be a government issue, or is this something in which the private sector should be involved? What are the potential benefits to private organizations that become involved in the business of space-debris removal?

Response: We have no particular expertise or involvement in space debris cleanup.

9) How can the U.S. government and the private sector work together to best utilize their combined resources?

Response: The B612 Foundation is pioneering the use of commercial procurement and management practices for interplanetary space mission deployment. We are interested in working with the US Government to explore ways to expand our innovative private-public partnership. The B612 Foundation has already put together one of the most experienced spacecraft development teams in the world, and has made significant technical progress on Sentinel. But we believe we can move even faster with if we were to work collaboratively with certain federal agencies. If resources were available and agreements could be worked out, we believe that launching and deploying Sentinel by 2016 is possible.

10) What can the U.S. government learn from the private sector's efforts to identify and characterize NEOs? Conversely, what can the private sector learn from the U.S. government's detection efforts?

Response: Much of the technology and know-how that will go into Sentinel is a result of government investment in space technology. It is crucial that the government continue to invest in such advanced technology.

11) The B612 Foundation's SENTINEL Mission is projected to discover more than 98% of all NEOs known to humanity during its six and a half year mission. What about the other 2%? What are the odds that a "continent destroyer" will be among those that evade detection?

Response: This question misstates the capabilities of Sentinel. The correct statement is that once Sentinel has completed its mapping of asteroids, we will have discovered about 1 million NEOs. And very nearly all of them (more than 98%) will have been discovered by Sentinel.

12) What is the current system for international coordination in the event of an imminent NEO threat? What recommendations do you have to improve that system?

Response: There is no yet agreed upon protocol for coordination of an imminent asteroid threat, although there is currently a process that has been proposed through the United Nations Committee on Peaceful Uses of Outer Space.

13) Do you know of any international private organizations that are involved in NEO detection or mitigation? If so, in what ways could they contribute to the combined detection and mitigation efforts of the U.S. government and private sector and foreign government?

Response: We do not know of any other private organizations (international or otherwise) besides the B612 Foundation with the plans or capability to discover and track asteroids at the scale of Sentinel. Given that there are a million NEOs that are "city killer" or larger, in order to make a significant dent in this problem requires a system capable of finding a hundred thousand or more asteroids per year. No other system besides Sentinel has this capability.

14) Have there been discussions and agreements on how much involvement in the mitigation process foreign governments are willing to provide to the U.S. government in the event of a NEO where destruction is limited to U.S. soil? How much aid is the U.S. willing to provide in a converse situation?

Response: We do not know of any such discussions. It should be noted though that if a large asteroid is found on a collision course with Earth, the likely cost of deflecting that asteroid would be insignificant compared to the potential loss should it be allowed to hit.

15) Considering the low probability of a devastating NEO impact, are detection and mitigation projects worth their high costs?

Response: A system to detect and track NEOs, which is the first step in protecting Earth from the hazard of asteroid impacts, is not expensive compared to the potential losses. Sentinel for instance has a budget which would be less than 1% of the current NASA budget. The benefits though are incalculable.

16) The President's Budget places NASA's asteroid strategy as a more visible component of the agency's mission, particularly in regard to human spaceflight. The agency is proposing combining agency efforts to ultimately have a human mission planned for 2021. What are your thoughts about the Administration's proposal? Specifically, can the various components NASA says it needs for a human mission benefit the overall goals we are discussing here today?

Response: The very small asteroids (approximately 25 feet across) that are potential targets of the proposed asteroid capture missions are too small to represent a threat to Earth.

Response to questions from Ranking Member Eddie Bernice Johnson:

1) This Committee is working on reauthorizing NASA for FY 2014 and beyond. In your view, what priorities with regard to NEOs do we need to address in legislation? How do we ensure that private-sector and international initiatives are effectively leveraged and integrated into a global response?

Response: We feel that a concerted effort must be made to find and track the roughly 1 million asteroids large enough to destroy a city. Less than 1% of such NEOs are currently tracked, which means that we are nearly certain to be blindsided by the next such object to hit Earth unless we do something about it. This is a chance to show American technical and scientific leadership because this truly is a global problem.

2) What are the challenges involved in assimilating NEO detection and characterization input from multiple observing platforms? How could this be done?

Response: Assimilating data from multiple observing problems is not an issue as this is already currently done. All observations of NEOs from telescopes around the world are funneled through the Minor Planet Center in Cambridge MA. Sentinel will also send its data to the Minor Planet Center.

3) In the past, NASA has experienced significant cost growth in several of its space science programs.
    a. What is the basis for your cost estimate for developing the Sentinel telescope and what gives you confidence that the estimate can be met?
    b. Has B612 requested an independent verification of the cost estimate?
    c. Do you have a "Plan B" if the cost becomes significantly greater than what is currently estimated?

Response: Sentinel is based on the designs of 2 previously flown spacecraft, the Kepler Space Telescope and the Spitzer infrared Space Telescope, both of which were built by our contractor Ball Aerospace. Because Ball has based the design of Sentinel on these successful programs, they were confident enough in their cost estimates to offer a firm fixed price proposal for Sentinel. In addition, B612 Foundation has completed a Program Concept and Implementation Review, in which the Sentinel project was reviewed by an independent review team (which included several NASA engineers and scientists). Because Sentinel is being built and managed under a fixed price contract, the risks of cost overruns are borne by Ball Aerospace. Appropriate reserves have been built into the cost of the contract.

4) You attribute B612's ability to use commercial contracting practices as the reason why it can develop Sentinel for less cost than NASA. B612 indicates that this can be done because requirements are stable and a firm-fixed price for the spacecraft can be used. Are there any technologies in Sentinel that require development, and if so, what is the risk of these encountering cost increases?

Response: The only new technology development being done for Sentinel involves its infrared imaging detectors. Before Ball issued its proposal for Sentinel, B612 Foundation funded a detector feasibility study between 2 competing detector subcontractors. On the basis of this work, both Ball and B612 agreed that the remaining detector work, while challenging, was an acceptable risk. And based upon this initial study, we were able to select a prime detector manufacturer while keeping the other manufacturer as a backup. To understand the risks even better, we have funded and built subscale detector prototypes, and have tested these prototypes under their planned operating temperature. We believe we understand the detector development risk. These risks and costs, as well as the appropriate reserves have been built into the main Sentinel contract.

5) Since B612 intends to provide free access of the data to the public, are there modifications to the data you will provide to NASA that form the basis for the Space Act Agreement established between the two parties? Would NASA need to enter into a data buy arrangement in addition to the SAA?

Response: We do not plan to modify the data provided to NASA and the public. The NASA Space Act Agreement with B612 Foundation stipulates the data that will be provided. If the federal government wishes to work more closely with B612 to accelerate the development of Sentinel, this will require federal investment, and data purchases are one of many possibilities for structuring such an arrangement.

Response to questions from Representative Steve Stockman:

1) What are the key technology demonstrations that would need to be conducted for deflecting asteroids?

Response: The scientific community agrees that asteroid deflection is possible with current technology, but it would certainly be advantageous to carry out a test mission on a non-threatening asteroid. But the key point is that all these technologies are useless against an asteroid we have not discovered yet. The very first priority must be to find and track asteroids.

2) What is the state of the readiness of the technology for the various methods of deflecting asteroids?

Response: The three principal methods of deflecting asteroids are kinetic impactors, gravity tractors, and nuclear standoff explosions. It would be feasible to test any of these technologies on a non-threatening asteroid.

3) What would be the effective range (in time/distance) of applicability of each different method of deflection?

Response: With many years of advance notice, deflecting an asteroid only requires a very tiny change in the asteroid trajectory to enable a successful deflection. With less than a few years notice, in general there are no known technologies for preventing an asteroid impact. That is why it is critical that asteroid search programs be carried out soon – to avoid the situation where we only have late notice and our only option is to evacuate the area and hope for the best.

Responses by Dr. Donald K. Yeomans
HOUSE COMMITTEE ON SCIENCE, SPACE, AND TECHNOLOGY

"Threats from Space: A Review of Private Sector Efforts to Track and Mitigate Asteroids and Meteors, Part II"

Questions for the Record, Dr. Donald K. Yeomans, Near-Earth Objects Program Office
Jet Propulsion Laboratory

Questions submitted by Rep. Steven Palazzo, Subcommittee on Space

1. Do we have the tools and technology necessary to detect Near Earth Objects (NEOs)? Once we identify an object, what are our means of tracking it?

    ANSWER: Since 1998, NASA has supported several ground-based optical telescope facilities for discovering and following-up NEOs. The progress for finding NEOs larger than one kilometer has been very impressive with a total discovery completion rate of more than 95 percent. Once a NEO discovery is made, a combination of professional and amateur astronomers provide the critically important follow-up optical observations that allow accurate orbits to be computed and the NEO's motion to then be accurately predicted for more than one hundred years into the future. Planetary radar observations, if available, are especially good for orbit refinement and for determining the NEO's size, shape and rotation characteristics. In addition, many amateur astronomers provide an observed time history of the NEO's ability to reflect light and hence, if these objects are irregularly shaped, these types of observations can be used to determine the rotation rate of the NEO. The Minor Planet Center (MPC) is the worldwide central node for receipt and distribution of NEO observation data. The MPC, which is funded by NASA and located at the Harvard-Smithsonian Astrophysical Observatory, collects and correlates NEO observation data from a variety of sources including amateur and professional astronomers for worldwide dissemination.

2. What categories of NEOs do we currently track, and which of these present potential cause for concern?

    ANSWER: NEOs are defined as those asteroids and comets that can approach the Earth's orbit to within about 30 million miles. In near-Earth space, asteroids outnumber comets one hundred to one. Of particular concern are the so-called potentially hazardous asteroids that can approach the Earth's orbit to within 5 million miles. Those near-Earth asteroids that are in Earth-like orbits about the sun are of most concern because they can repeatedly approach the Earth. Currently, all discovered NEOs and the subset of potentially hazardous asteroids are being tracked with ground-based optical telescopes to ensure that enough observations are available to confidently predict their orbital paths.

3. Both Dr. Holdren and Administrator Bolden testified to our committee in March that we have a long way to go to accomplish the goals established by Congress in the NASA Authorization Act of 2005 of detecting 90 percent of the NEOs with a diameter of 140 meters or greater by 2020. What are the most important steps that should be taken in the next five years to accomplish these goals?

ANSWER: The total population of NEOs with diameters 140 meters and larger is thought to be about 20,000 and NASA-supported surveys have discovered about 25 percent of this population. It is very unlikely that the existing ground-based optical surveys will reach 90 percent completion at this size range by 2020, but NASA is evaluating systems that could make it possible by 2030 and will implement enhancements and acquire the required additional capability as soon as we are able to do so. The B612 Foundation plans for Ball Aerospace to build an infrared space-based NEO discovery telescope. This type of telescope would efficiently capture the heat (infrared re-radiated sunlight) from dark asteroids and could do so without the interruptions due to weather and daylight that affect ground-based assets. The B612 Foundation plans to philanthropically fund this effort and NASA has signed a Space Act Agreement with B612 to provide advisory information as well as spacecraft tracking and navigation support. In addition, NASA's Human Exploration and Operations Mission Directorate and Science Mission Directorate, through the Joint Robotic Precursor Activity office, are studying instrument concepts for a mission of opportunity to be hosted on a US Government or commercial spacecraft in geosynchronous orbit that will be capable of detecting and tracking asteroids in orbits very similar to Earth's; NASA released a Request for Information (RFI) in August 2012 and is studying the instrument concepts that were submitted.

4. Can you describe the potential range of damage caused by impacts from NEOs?

ANSWER: There are vastly more small NEOs than large ones, so the most likely impact will be due to a relatively small NEO. At the small end of the NEO population size range, NEOs with diameters of 20-30 meters would be expected to cause air blasts that could cause local destruction of property and injuries (including possible fatalities) if they were to impact over populated areas. The Chelyabinsk Russia air blast of February 15, 2013, was caused by a near-Earth asteroid approximately 20 meters in size and the more powerful Tunguska blast over Russian Siberia in 1908 was due to an object of about 30-50 meters in diameter. Impacting NEOs of about 140 meters in diameter would cause regional devastation over land and possibly cause tsunamis in the more likely event they impacted into the oceans. NEOs larger than a kilometer or two would be expected to cause catastrophic effects for all nations, but especially in third world countries that lack the resources to recover from extensive crop failures and widespread infrastructure damage.

5. What are the areas in which there is a lack of knowledge or understanding of NEOs? And what barriers does the private sector face in gaining the knowledge necessary to quantify and mitigate the risk of NEO impacts?

ANSWER: In terms of the effects caused by the impacts of relatively large NEOs, the expected damage due to water impacts and the subsequent generation of tsunamis is an area of great uncertainty. Given that the Earth's surface is about two thirds covered with oceans, an ocean impact is the most likely scenario but the efficiency with which a NEO impact could cause a tsunami is not well understood.

Deflection techniques are another area in need of further study. While considerable thinking has gone into a variety of approaches (including a purposeful collision by a high speed spacecraft as well as redirection of the NEO's path by use of thrusting), more extensive analysis will be required before any of them can be considered well understood.

The success with which the current surveys have undertaken the NEO discovery searches, as well as the success of the subsequent follow-up observations and characterization studies, is largely due to the cooperative efforts of NASA and several diverse entities within the academic community. By means of NASA's annual peer-review proposal process, funding is provided to the most promising, innovative and successful NEO researchers. It is difficult to think of a more efficient process for exploiting the widespread expertise and knowledge centers that are required for addressing the complex NEO issues.

6. From where you operate, how would you describe the level of coordination between governments and outside organizations? What improvements need to be made?

ANSWER: Under the auspices of the United Nations Committee on the Peaceful Uses of Outer Space (UN COPUOS), Scientific and Technical Subcommittee, there has been an ongoing effort by the Subcommittee's NEO Working Group to provide an international framework for the detection and warning for NEOs that may represent a threat to Earth. An International Asteroid Warning Network (IAWN) has been proposed to link together the institutions that are already performing many of the proposed functions of the IAWN. Many of these functions are already being successfully carried out by NASA sponsored efforts. In addition, a Space Mission Planning Advisory Group (SMPAG) has been proposed to facilitate the gathering of necessary NEO data and to coordinate among the international entities that would likely be involved in mitigation and civil defense activities. While improvement to coordination efforts among international partners should continue, there has been an excellent start to these activities. There is reason to believe that the ongoing process will be successful in providing international guidelines and protocols that will guide a future international response to a threatening NEO.

7. What can the U.S. government do to encourage the advancement of private sector technologies that detect and track NEOs?

ANSWER: NASA support and funding has been key to the success of the search and post-discovery follow-up and characterization observations of NEOs. Without this government support, none of this success would have been possible. As an example, NASA is working collaboratively with the privately funded B612 Foundation by providing technical assistance and operational support through a Space Act Agreement on their efforts to build a space observatory to detect 100-meter size objects and larger that could come near Earth's orbit. NASA will also provide B612 access to our Deep Space Network for telecommunications with the spacecraft for commanding and data downlink.

However, the private sector also has a very important role to play. The sophisticated telescopes, sensors, computers and other technologies used to provide and analyze the data taken on NEOs are largely a result of work done by the private sector. By providing the necessary support to NASA's NEO Observations Program, the best peer-reviewed innovative ideas are funded to discover, follow-up and physically characterize the population of NEOs. The recipients of these successful proposal grants often turn to the private sector to provide the technology to carry out their innovative ideas. This process insures a steady stream of new ideas that push the private sector providers to advance their technologies and remain competitive.

8. Should the cleanup of space debris primarily be a government issue, or is this something in which the private sector should be involved? What are the potential benefits to private organizations that become involved in the business of space-debris removal?

    ANSWER: NASA, with the help of DoD, industry, and academia, has completed an extensive review of orbital debris removal concepts. None of these concepts currently meet minimum requirements for technical maturity and affordability. However, as directed by the President's National Space Policy (2010), NASA and DoD are continuing to pursue development of early-stage technologies and techniques to mitigate and remove on-orbit debris, reduce hazards, and increase understanding of the current and future debris environment.

    Any remediation of the near-Earth space environment, if and when it happens, will necessarily involve an international effort. International treaties prevent a country from removing space objects which do not belong to it and the U.S. is responsible for less than one-third of all cataloged debris now in Earth orbit. In fact, only 6% of all objects now in low Earth orbit with a mass larger than one metric ton (these are the objects with highest potential for causing future damage) belong to the U.S.

    Early-stage technology work on debris removal technologies has begun and will likely continue in order to develop the capabilities necessary in time for potential future operations. Unlike the recycling of waste on Earth, orbital debris does not yet have an intrinsic value which would support a purely commercial undertaking. Efforts to date have been conducted by national governments, although the capabilities of the private sector could be leveraged in the future.

9. How can the U.S. government and the private sector work together to best utilize their combined resources?

    ANSWER: As noted in the response to question 7, the NASA supported NEO Observations program is already working closely with the private sector to bring into use the most sophisticated technologies. The U.S. space program has always been a driving force for private sector technology innovation.

10. What is the current system for international coordination in the event of an imminent NEO threat? What recommendations do you have to improve that system?

    ANSWER: As noted in the response to question 6, there is a successful, ongoing international process to coordinate the international response to a NEO threat, either an imminent threat or one that is years into the future.

11. Do you know of any international private organizations that are involved in NEO detection or mitigation? If so, in what ways could they contribute to the combined detection and mitigation efforts of the U.S. government and private sector and foreign governments?

    ANSWER: There are several private organizations already involved in NEO detection and mitigation in cooperation with NASA the European Space Agency, and the European Commission. For example, in the U.S., there are plans for Ball Aerospace to build an infrared space-based NEO discovery telescope for the B612 Foundation. The B612 Foundation plans to philanthropically fund this effort and NASA has signed a Space Act Agreement with B612 to provide advisory information as well as spacecraft tracking and navigation support. Dr. Ed Lu has testified concerning this concept. In Europe, the Astrium Company is working with NEOShield, currently funded by a European Commission grant, to make plans for a NEO deflection demonstration mission as well as providing technical support for a number of NEO deflection studies being undertaken by NEOShield. NASA personnel are already involved with these NEO efforts and NASA will continue to look for ways to leverage the expertise and resources of the private sector and foreign governments.

12. Have there been discussions and agreements on how much involvement in the mitigation process foreign governments are willing to provide to the U.S. government in the event of a NEO where destruction is limited to U.S. soil? How much aid is the U.S. willing to provide in a converse situation?

    ANSWER: Mitigation of an impact is an international issue and will require a cooperative response by all space-faring nations. While there have been preliminary discussions about the mitigation process among space-faring nations under the auspices of UN COPUOS, these discussions are ongoing and have not reached the level of detail required to address the issue of which nation, or nations, would be authorized to attempt a NEO deflection mission. In addition, there have not yet been substantive discussions on how the necessary resources would be provided for a NEO mitigation campaign. These issues will be addressed in future meetings of UN COPUOS or among technically capable nations.

13. How would you characterize the U.S. government's participation in UN COPUOS as it relates to asteroid detection and disaster mitigation?

ANSWER: As noted in the response to question 6, the U.S. government, with NASA in the lead, is currently participating in a productive, ongoing effort within UN COPUOS to establish an International Asteroid Warning Network (IAWN) and a Space Mission Planning Advisory Group (SMPAG). While the vast majority of the NEO discoveries, follow-up and characterization observations to date have been carried out by NASA, the European Space Agency, Japan, Russia and other nations are becoming more engaged and these cooperative efforts are being encouraged as a result of the UN COPUOS activities.

14. NEO detection and mitigation is an international concern. How are foreign governments contributing to the costs of U.S. government NEO detection technologies, and vice versa? For example, what is the U.S. government's involvement in the Near Earth Object Dynamic Site compared to that of foreign governments, and is this an adequate distribution of cost sharing and responsibility?

ANSWER: There has been no direct funding exchanged between NASA and its international partners for NEO activities, but there have been significant cooperative efforts and information exchanges between these partners. NASA is currently providing the vast majority of resources for the discovery, follow-up and physical characterization of NEOs through support of projects at US institutions. NASA-supported projects have discovered approximately 98% of all NEOs. However, the European Space Agency's (ESA) Space Situational Awareness (SSA) Program is now funding a one-meter telescope in Tenerife that provides valuable follow-up observations and ESA also completely funds the Near-Earth Object Dynamic Site (NEODyS) in Pisa, Italy. This latter site presents the results of orbital computations and, if appropriate, impact probability calculations for all discovered NEOs and it provides a valuable cross check on the parallel efforts being carried out by the NASA-supported NEO Program Office located at the Jet Propulsion Laboratory. The JPL and Pisa offices are in constant communication to verify their respective orbital and impact probability computations. In those rare cases when an object has a non-zero impact probability in the near future, JPL and Pisa cross check one another before posting results on their respective websites.

ESA's SSA program is also funding a NEO data collection activity in Germany where all the physical characteristics for known near-Earth objects are archived and made available to the international community via their website. The NEOShield effort, funded by the European Commission and noted in the response to question 11, is also active in studying the various techniques that could be used to deflect or disrupt a NEO on an Earth threatening trajectory. Japan's very successful Hayabusa mission achieved a rendezvous with near-Earth object Itokawa in 2005 and returned a small sample from the surface of this object in June 2010. NASA provided some spacecraft tracking and spacecraft navigation support and several U.S. members of the scientific community participated on the Hayabusa Science Team. Japan is currently planning a Hayabusa 2 mission to another near-Earth asteroid and there will likely be continued cooperative efforts with NASA participation. These Hayabusa NEO rendezvous missions, along with NASA's rendezvous mission to near-Earth asteroid Eros in 2000, the rendezvous and sample return mission OSIRIS-REx, and the planned first-ever mission to identify, capture, and redirect an asteroid into orbit around the Moon for future astronaut rendezvous and sampling are, and will be, important sources for the detailed understanding of near-Earth object structures and compositions. Russia is also undertaking studies to better understand the optimal techniques for deflecting near- Earth asteroids.

15. Considering the low probability of a devastating NEO impact, are detection and mitigation projects worth their high costs?

    ANSWER: An early discovery of an Earth-threatening NEO would allow the time to safely deflect it with existing technologies. As the search for NEOs continues, and more and more of them are discovered and tracked one hundred or more years into the future, their risks to Earth can be evaluated. More than 95 percent of the largest NEOs (1 kilometer and larger) have already been discovered for a total cost of less than $70 million spread over 15 years, and it is reassuring to know that none represent a serious impact threat in the next one hundred years. Despite the low probability of a devastating NEO impact, it seems prudent to continue to invest in strategies for early warning and mitigation.

16. The President's Budget places NASA's asteroid strategy as a more visible component of the agency's mission, particularly in regard to human spaceflight. The agency is proposing combining agency efforts to ultimately have a human mission planned for 2021. What are your thoughts about the Administration's proposal? Specifically, can the various components NASA says it needs for a human mission benefit the overall goals we are discussing here today?

    ANSWER: The FY 2014 President's budget contains funding to accelerate technology development in areas important in their own ways for exploration, including advanced solar propulsion. It also provides funds to begin planning for an Asteroid Redirect Mission (ARM) that would utilize advanced solar propulsion technologies to rendezvous with, and redirect, a small NEO (about 7 meters in diameter) and bring it back into a stable orbit in the lunar vicinity for sampling by astronauts. While the ARM planning is not driven by science objectives, the necessary search for a suitable target asteroid would certainly provide an increase in the discovery rate of NEOs along with a concomitant increase in the characterization of this population. Since the stringent requirements for a suitable ARM target body dictate that these targets be in rather Earth-like orbits, this population of Earth's closest celestial neighbors would be better characterized than is currently the case. Those NEOs that are most easily reached by spacecraft, or most suitable for round trip human exploration, are the same objects that represent the greatest likelihood of striking Earth. A better understanding of this population would benefit both planetary science and planetary defense.

Questions from Ranking Member Eddie Bernice Johnson to Dr. Yeomans

1. This Committee is working on reauthorizing NASA for FY 2014 and beyond. In your view, what priorities with regard to NEOs do we need to address in legislation? How do we ensure that private-sector and international initiatives are effectively leveraged and integrated into a global response?

   ANSWER: To best understand and characterize specific NEOs, early detection is key. The approach within the NEO Program in the President's FY14 budget request is to expand the existing NEO detection and characterization activities. This includes making available more time on existing ground-based observatories capable of detecting or characterizing NEOs, such as Pan-STARRS or the Space Surveillance Telescope (SST).

   Also, the President's budget contains funding for an Asteroid Redirect Mission (ARM) that would utilize advanced solar propulsion technologies to rendezvous with, and redirect, a small NEO (about 7 meters in diameter) and bring it back into a stable orbit in the lunar vicinity for possible study by astronauts. While the ARM planning is not driven by science objectives, the necessary search for a suitable target asteroid would certainly provide an increase in the discovery rate of NEOs along with a simultaneous increase in the characterization of this population. Since the stringent requirements for a suitable ARM target body dictate that these targets be in rather Earth-like orbits, this population of Earth's closest celestial neighbors would be better characterized than is currently the case. A better understanding of this population would benefit both planetary science and planetary defense.

   Deflection techniques are another area in need of further study. While considerable thinking has gone into a variety of approaches (including a purposeful collision by a high speed spacecraft as well as redirection of the NEO's path by use of thrusting), more extensive analysis will be required before any of them can be considered well understood. The on-going international cooperative mission studies and NEO data gathering are effective ways to leverage the resources and expertise in the international community.

2. What are the risks, if any, of relying on non-government organizations to provide data needed to meet a congressional mandate? If such non-governmental capabilities are delayed or become unavailable, what options would the government have to obtain the needed data?

   ANSWER: NASA has established a Space Act Agreement with the B612 Foundation on their Sentinel Project to provide some technical advice, spacecraft tracking and navigation services as well as processing the observations through the existing NASA-supported Minor Planet Center and NASA's Near-Earth Object Program Office. While NASA fully supports the Sentinel activity, NASA continues to monitor its progress and assess its viability, and more robust alternative options are being studied. In addition, NASA's Human Exploration and Operations Mission Directorate (HEOMD) and Science Mission Directorate (SMD), through the Joint Robotic Precursor Activity (JRPA) office, are studying instrument concepts for a mission of opportunity to be hosted on a US Government or commercial spacecraft in geosynchronous orbit that will be capable of detecting and tracking asteroids in orbits very similar to Earth's; NASA released a Request for Information (RFI) in August 2012 and is studying the instrument concepts that were submitted. Another alternative would be to place a space-based infrared telescope at a location about one million miles on the sunward side of the Earth at the so-called L1 point. This would allow continuous observations and image downlinks of near-Earth objects down to sizes below 100 meters. NASA is committed to satisfying the Congressional mandate to find and track 90% of the near–Earth asteroids larger than 140 meters that are Earth impact hazards.

3. What are the key challenges to meeting Congressional direction on surveying, detecting, and characterizing near-Earth objects equal to or greater than 140 meters in diameter by 2020? In your opinion, is this a technological issue or are budgetary resources the key pacing item?

ANSWER: The key challenge in meeting the Congressional direction outlined in the 2005 NASA Authorization Act is to efficiently detect the remaining thousands of undiscovered 140 meter and larger sized NEOs within the next several years.

While ground-based surveys are making excellent progress increasing the discovery rate, it will require a space-based infrared NEO telescope to significantly increase the current detection rate. The expertise and technology exist to build, launch and operate an infrared telescope, located either in a heliocentric orbit at a distance similar to that of the planet Venus or located on the Earth-Sun line about one million miles sunward of Earth (at the so-called first Lagrange point, or L1). The B612 Foundation has announced plans to philanthropically fund an effort to operate an infrared telescope in a Venus-like orbit. NASA has signed a Space Act Agreement with B612 to provide advisory information as well as spacecraft tracking and navigation support. Dr. Ed Lu has testified concerning this concept. NASA has also funded an advanced infrared detector development that could be employed on an infrared telescope operating at the Sun-Earth L1 position. In addition, NASA is advancing work on instrument concepts for a mission of opportunity to be hosted on a US Government or commercial spacecraft in geosynchronous orbit that will be capable of detecting and tracking asteroids in orbits very similar to Earth's. NASA is also evaluating reactivating our NEOWISE activity, a very successful use of an Earth-orbiting, Wide-field Infrared Survey Explorer (WISE) space telescope that was used in 2010-2011 to find and physically characterize near-Earth objects.

4. What are the challenges involved in assimilating NEO detection and characterization input from multiple observing platforms? How could this be done?

ANSWER: NEO detection, follow-up observations and physical characterization measurements are already being carried out at multiple international observatories and coordinated within a few NASA and ESA supported facilities. The discovery and follow-up observations are forwarded to the international central clearing house at the NASA-supported Minor Planet Center (MPC) and from there these observation are forwarded to both the computational centers at the Jet Propulsion Laboratory (JPL) and the NEODyS facility in Pisa, Italy. This latter facility is supported by the European Space Agency's Space Situational Awareness program. In turn, the MPC, JPL and the NEODyS facilities provide observing predictions (ephemerides) to the observer community. At JPL, observing position, velocity and range predictions are provided to observers at the planetary radar facilities at Goldstone's 70-meter antenna in southern California and to observers at the 305-meter antenna located in Arecibo, Puerto Rico. JPL's automated Horizons on-line ephemeris service provides more than 80,000 ephemeris products daily to the international research communities. While challenges will arise for assimilating NEO detection and characterization inputs from multiple ground-based and space-based platforms, the successful track record for ground-based observing platforms suggests these challenges will be met in a straightforward manner.

5. To what extent do international space agencies or international facilities contribute to NASA's NEO survey and/or a worldwide effort of surveying, tracking, and characterizing potentially hazardous near-Earth objects? How effective is communication and data-sharing on near-Earth object tracking among nations? What is the degree of international involvement in studying deflection options for NEOs? What more could be done?

ANSWER: The international community of NEO researchers is well coordinated and has been working cooperatively for several years. The international communication and data sharing channels are operating well.

While NASA has been responsible for almost all of the NEO discoveries and the efforts to physically characterize a representative sample of these objects, there has been recent progress in NEO research in the international community as well. In particular, there are ongoing efforts, funded by the European Space Agency, to collect and archive the existing physical data on NEOs, to compute orbits and predict close Earth approaches in parallel with similar efforts at the Jet Propulsion Laboratory and to provide the critically important post-discovery follow-up observations that allow accurate orbits to be determined for recently discovered NEOs. Under a grant provided by the European Commission, the NEOShield program is carrying out studies to determine the optimal techniques for NEO deflection and as well as outlining plans for an asteroid deflection demonstration mission. The Japanese Aerospace Exploration Agency (JAXA) has successfully carried out the rendezvous mission Hayabusa to near-Earth asteroid Itokawa and returned a small sample to Earth in June 2010 so that the international community can carry out detailed analyses to determine the elemental composition of this NEO. JAXA is also making plans for a Hayabusa 2 mission to rendezvous with, and return a sample from, another near-Earth asteroid. U.S. scientists are interacting or participating with the ESA, European Commission and JAXA activities. In addition, Russian scientists have carried out a number of studies to investigate options for deflecting NEOs.

All international discovery and follow-up observations as well as preliminary orbital computations for NEOs are carried out at the NASA-sponsored Minor Planet Center (MPC) in Cambridge Massachusetts. The MPC is the recognized international clearinghouse for these data and its work is carried out under the auspices of the International Astronomical Union. The MPC collects and archives these data, notifies the international observing community which objects need follow-up observations and is the first agency to announce the possibility of short term Earth close approaches or impacts. The MPC has been in continuous operation since 1947 and is instrumental in the effective and timely communication of NEO discoveries, coming close Earth approaches and future NEO observing opportunities.

Because the first step in deflecting hazardous NEOs is to find them early enough to allow a successful mitigation campaign, the emphasis to date has appropriately been on the NEO search, follow-up and characterization efforts. Nevertheless, there have been several studies carried out to investigate the optimal techniques and mission designs to deflect NEOs. For example, JPL personnel carried out a 2012 study to better understand the viable mission options for deflecting a NEO designated 2011 AG5 in the unlikely event that this object's non-zero impact probability in 2040 remained a possibility. While recent observations of this object allowed its orbit to be refined to such an extent that the 2040 Earth impact possibility was ruled out, the study outlined the steps necessary for planning, designing and launching an impacting spacecraft to deflect a hazardous NEO.

Finally, under the auspices of the United Nations Committee on the Peaceful Uses of Outer Space (UN COPUOS), Scientific and Technical Subcommittee, there has been an ongoing effort by the Subcommittee's NEO Working Group to provide an international framework for the detection and warning for NEOs that may represent a threat to Earth. An International Asteroid Warning Network (IAWN) has been proposed to link together the institutions that are already performing many of the proposed functions of the IAWN. Many of these functions are already being successfully carried out by NASA sponsored efforts. In addition, a Space Mission Planning Advisory Group (SMPAG) has been proposed to facilitate the gathering of necessary NEO data and to coordinate among the international entities that would likely be involved in mitigation and civil defense activities. While improvement to coordination efforts among international partners should continue, there has been an excellent start to these activities. There is reason to believe that the ongoing process will be successful in providing international guidelines and protocols that will guide a future international response to a threatening NEO.

HOUSE COMMITTEE ON SCIENCE, SPACE, AND TECHNOLOGY

"Threats from Space: A Review of Private Sector Efforts to Track and Mitigate Asteroids and Meteors, Part II"

Questions for the Record, Dr. Donald K. Yeomans, Near-Earth Objects Program Office
Jet Propulsion Laboratory

Questions submitted by Rep. Steve Stockman

- What are the key technology demonstrations that would need to be conducted for deflecting asteroids?

  ANSWER: Deflection techniques are an area in need of further study. While considerable thinking has gone into a variety of approaches, including a purposeful collision by a high speed spacecraft as well as redirection of the NEO's path by use of thrusting. The proposed Asteroid Redirect Mission, which intends to use advanced solar electric propulsion technologies to rendezvous with and redirect a small NEO, could also inform potential asteroid deflection techniques. Regardless the chosen method, more extensive analysis will be required before any of them can be considered well understood.

- What is the state of the readiness of the technology for the various methods of deflecting asteroids?

  ANSWER: The most effective approach to mitigation of a potential asteroid impact threat is highly dependent on the scenario. Near-term impact of an asteroid tens of meters in diameter requires a significantly different approach than the threat of a larger object that might impact decades in the future. The orbit parameters of the potential impactor are also a significant factor in determining an effective mitigation strategy. Therefore, a "toolkit" of mitigation approaches needs to be developed at the conceptual level to address the range of potential impact threats. While considerable thinking has gone into a variety of approaches (including a purposeful collision by a high speed spacecraft as well as redirection of the NEO's path by use of thrusters), more extensive analysis will be required before any of them can be considered well understood. As a next step, in FY 2014, NASA's Office of the Chief Technologist plans to develop a roadmap of mitigation technologies.

- What would be the effective range (in time/distance) of applicability of each different method of deflection?

  ANSWER: The most effective deflection methods are not carried out when an impacting object is on its final trajectory with only a few days or even weeks before impact, but rather several years and many orbits of the Sun away from the predicted time of impact. The most important element of any deflection method is to find the hazardous object as early as possible. Each NEO deflection campaign would depend upon several variables including the size, mass, rotation and composition of the NEO as well as the time available before impact.

Responses by Dr. Michael F. A'Hearn

Threats from Space: A Review of Private Sector Efforts to Track and Mitigate Asteroids and Meteors, Part II

Wednesday, April 10, 2013

House of Representatives

Committee on Science, Space, and Technology

Response to Additional Questions

Michael F. A'Hearn

Questions from Rep. Steven Palazzo

1. Do we have the tools and technology necessary to detect Near Earth Objects (NEOs)? Once we identify an object, what are our means of tracking it?

We certainly have the tools to discover the larger NEOs and tools to discover ones down to 140m are under development. We have a fully functioning system to collect observations and calculate orbits in order to determine which objects are NEOs and which of those are likely to be threatening. Tracking the objects after discovery, in order to determine reliable orbits, is easily done with relatively small telescopes and many observers participate in this activity.

There are only two tools we will be lacking if all the planned facilities are completed are 1) a tool to detect "all" (realistically 90%) of the NEOs smaller than 140m in diameter and down to the smallest ones that do significant damage, somewhere around 30-50 m in diameter, and 2) a tool to detect long period comets sufficiently far in advance to react.

2. What categories of NEOs do we currently track, and which of these present possible cause for concern?

We currently track all discovered NEOs except the very smallest ones that are too faint to be tracked and most of which are too small to cause significant damage on Earth. Of the ones we do track, most have been shown to present no hazard to Earth for a century or more. There is, however, a subset of such objects that have frequent close passages by Earth. If one of those passes the Earth through a relatively small volume of space (smaller than the accuracy of our predictions), the NEO can then be deflected onto an orbit that does impact Earth years or decades later.

3. Both Dr. Holdren and Administrator Bolden testified to our committee in March that we have a long way to go to accomplish the goals established by Congress in the NASA Authorization Act of 2005 of detecting 90 percent of the NEOs with a diameter of 140 meters or greater by 2020. What are the most important steps that should be taken in the next five years to accomplish these goals?

As far as I know, it is no longer possible to achieve the goal of discovering 90% of the impactors down to a diameter of 140m by 2020. The facilities that could do this on reasonable time scales include 1) the Large Synoptic Survey Telescope, which is funded primarily for other science through NSF and DOE funding together with private partnerships, but which is not schedule to begin operations until 2021 or so, 2) the Sentinel System, about which we have heard in this hearing and which expects to be able to complete the survey by about 2023 or 2024 (launch in 2017 or 2018 and 6 ¾ years of surveying), 3) a satellite based on NASA's WISE satellite with techniques developed under the NEOWISE program, which was proposed (but which did not win in a purely scientific competition) to be placed at the L1 Lagrangian point between Sun and Earth, and 4) the Space Surveillance Telescope discussed by Dr. Yeomans and funded by DARPA and the U. S. Air Force, and about which I have no information.

As a practical matter of development time and survey time, I doubt that there is anything that could be done to achieve the goal by 2020, although completion on slightly shorter time scales than currently planned is probably possible with significantly accelerated funding.

4. Can you describe the potential range of damage caused by impacts from NEOs?

NEOs are capable of causing every imaginable type of damage from none at all up to a global catastrophe resulting in the extinction of many species, including humankind. The most common events will be events like the one that occurred in Chelyabinsk, Russia on February 15 of this year. More than 100 people were injured, mostly by flying glass, and one building collapsed. Such events should occur every several decades, although some will occur over uninhabited areas, including the oceans. NEOs greater than 1-2 km in diameter can cause global devastation, including dramatic climatic change and extinction of species. If we exclude the long-period comets, we have discovered >90% of these objects and shown that none of the discovered ones present a threat within the next century or more. These events, including the very infrequent impacts of long period comets, are likely to occur on average only once every 100 million years. An event that occurs once a century would be one that could wipe out a city if the impact occurred over or close to the city.

5. What are the areas in which there is a lack of knowledge or understanding of NEOs? And what barriers does the private sector face in gaining the knowledge necessary to quantify and mitigate the risk of NEO impacts?

The most significant gap in our understanding of NEOs is in their internal structure and composition. These parameters have a large effect on the amount of damage that would be caused by an impact and also a large effect on the effectiveness of any mitigation technique, whether aimed at altering the NEO's trajectory or at destruction of the NEO.

I am not aware that the private sector faces any more obstacles than does NASA in gaining the knowledge, at least in principle. Most of NASA's satellites and launch vehicles and even ground-based telescopes are developed by private industry. The characterization knowledge is generally public. The only area in which private industry would face an obstacle is in simulating in detail the amount of damage from an impact, since most of the computer codes that one would use for this are also used to simulate weapons tests and are therefore under tightly restricted access. Private industry might face practical issues of scale of operations and financing.

6. From where you operate, how would you describe the level of coordination between governments and outside organizations? What improvements need to be made?

As one who works with NASA primarily on the scientific side rather than on the NEO-hazard side, it appears to me that the coordination between the government and outside organizations regarding the NEO hazard is working well.

7. What can the U. S. government do to encourage the advancement of private sector technologies that detect and track NEOs?

In my personal view, encouragement of private sector technologies should be limited to the sharing of large resources, of which only a small fraction of the resource is needed by the private venture. The Sentinel project, for example, will downlink its data to Earth using NASA's Deep Space Network (DSN) but the project needs only a very small fraction of the total capacity of the DSN, which NASA uses to communicate on a regular basis with a very large number of spacecraft. Providing a small portion of the DSN, at no cost or even at a low pro-rated cost, represents a huge cost savings for a private venture that could not realistically build a complete downlink system at reasonable cost unless the spacecraft were very close to Earth. Any similar capabilities of the government that are expensive because of the scale of the operation, and of which the private venture needs only a small fraction, could be provided in a similar way as an encouragement.

8. Should the cleanup of space debris primarily be a government issue, or is this something in which the private sector should be involved? What are the potential benefits to private organizations that become involved in the business of space-debris removal?

The existence of space debris is a consequence of both governmental and private activities, and the governmental activities have been by many different countries and organizations. Similarly, both government and private industry benefit from the removal of space debris. The entire population of Earth will benefit from removing large pieces of space debris in low-Earth orbit (LEO), *i.e.*, debris that will undergo atmospheric drag and plunge to Earth. Removing space debris is a task that can certainly be carried out by private industry provided they can find a profit in doing so. Thus it appears to me that there is a role for private industry both in performing the clearance of space debris and in the funding of that effort, but there is also a role in the funding for governments around the world.

9. What is the current system for international coordination in the event of an imminent NEO threat? What recommendations do you have to improve that system?

Currently all international coordination is through the UN's Committee on Peaceful Uses of Outer Space (COPUOS). Certainly the UN is key in ensuring that all countries are made aware of the situation. I think that multi-lateral coordination directly among the countries capable of playing a major role is the next essential step. Ideally, the multilateral discussions would include a COPUOS representative to ensure that the same information is available to everyone.

10. Do you know of any international private organizations that are involved in NEO detection or mitigation? If so, in what ways could they contribute to the combined detection and mitigation efforts of the U. S. government and private sector and foreign governments?

There are many groups of amateur astronomers around the world, many using remotely operated telescopes far from their homes, who have professional class telescopes. Although they do not survey to first discover NEOs, they do discover some. More importantly, they provide the very rapid follow-up observations that are necessary to determine the orbits of newly discovered asteroids. Without these amateurs and the smaller number of professionals doing follow-up, many of the newly discovered asteroids would be lost and we would never know whether they are hazardous NEOs or not. These groups are already well integrated into the process through the International Astronomical Union's Minor Planet Center (MPC, physically located at the Harvard-Smithsonian Center for Astrophysics but funded entirely by NASA).

Internationally most of the survey projects are, at least in part, governmentally funded in one way or another. The satellite recently launched by Canada (NEOSAT) and the one in development in Germany are both government funded. Ground-based surveys may be funded in part by universities or other entities but in most cases the funding is ultimately derived from the national governments.

Thus the only private initiatives of which I am aware are the domestic B612 Foundation's Sentinel project, the mostly domestic LSST Corporation's LSST, and this loose collaboration of amateurs. There may be some additional private organizations, , but if so I am not aware of them.

11. Given that mitigation efforts for an NEO impact would be determined on a case-by-case basis, can you discuss dome of the more general, across-the-board mitigation strategies that both private entities and governments could implement?

In our NRC study, we assumed that mitigation would be undertaken using the least disruptive technique capable of mitigating the hazard. With that assumption, there are only four classes of techniques for mitigation. For the smallest impactors, say up to 30-50 meters (up to Tunguska-class events), evacuation is the simplest mitigation against loss of life, although this does not mitigate against property damage. For somewhat larger impactors, say up to 200 meters, provided one has several decades of advance warning, slow push/pull techniques can change the orbit of an NEO enough to miss Earth. Many different techniques have been proposed to slowly push or pull a small NEO, but the most studied technique, which is also among the most independent of the physical properties of the NEO, is the gravity tractor. Since the gravity tractor can only apply a very small force to pull the NEO, it must operate for years to decades in order to change the orbit of the NEO sufficiently. For a subset of NEOs that have repeated close encounters with Earth, it is often sufficient to change the orbit just enough to avoid a certain space near-Earth at a future encounter (a space often termed a keyhole) so that one does not need to act quite as far in advance. Moving up the scale, a kinetic impactor can change the NEO's orbit enough to miss Earth for small NEOs with only years of operation or NEOs up to a couple of km with decades of operation. For larger NEOs, the only known technique that is sufficiently powerful is a nuclear weapon, best executed as a standoff blast that vaporizes one side of the NEO to push it in the opposite direction. More details are provided in my response to Representative Stockman.

12. Considering the low probability of a devastating NEO impact, are detection and mitigation projects worth their high costs?

The low probability of a devastating impact certainly raises the question of how much one should spend on this effort and one's attitude toward risks varies dramatically from one person to another. What sets these events apart from most other natural disasters is the fact that we know, at least in principle and to a large extent in practice, how to predict AND PREVENT the event. To illustrate the probability, an event similar to Chelyabinsk, with more than 100 people injured (primarily by flying glass) but nobody killed, should occur every several decades or several times a century. An event like Tunguska, which flattened every tree over 2000 km^2, should occur ever 1-2 centuries. This is devastating at the local level but not at the national or global level.

Because the likelihood of an event is low, and since people's tolerance of risks varies dramatically, the extent to which the government should address the issue is basically a political decision on the degree of risk aversion or risk tolerance of the American people. I could only report on my personal sense of risk, which is quite likely not representative. I am personally sufficiently risk tolerant that I would not sacrifice NASA's science or exploration programs, but would be happy to see the effort funded as an increment to NASA's budget. This is consistent with the phraseology in the NRC report, which represented a consensus of the steering committee of the NRC committee.

13. The President's Budget places NASA's asteroid strategy as a more visible component of the agency's mission, particularly in regard to human spaceflight. The agency is proposing combining agency efforts to ultimately have a human mission planned for 2021. What are your thoughts about the Administration's proposal. Specifically, can the various components NASA says it needs for a human mission benefit the overall goals we are discussing here today?

I am very disappointed in the current plans of the administration for NASA. For as long as I can remember, the Science and Human Exploration Divisions set their own priorities with good interactions when synergy between the two made sense (*e.g.*, scientific instruments/experiments on the shuttle and on the space station; return of samples from the moon for scientific analysis) but the bulk of the two programs was set by scientific goals on the one hand and human exploration goals on the other. The administration's current plan appears to subordinate planetary science to human exploration, meaning that the best science is likely to be excluded. The president's budget also goes contrary to the recommendations of the NRC's recent decadal survey of planetary science. The decadal survey placed goals in priority order – 1. Maintain the scientific excellence in the community, b) bring the Discovery Program back up to the frequent launches for which it was created with a new Announcement of Opportunity at least every two years, c) bring the New Frontiers Program to its planned level (it is close to the planned level now), d) keep the flagship missions up to their planned level of 1 per decade. Regrettably, the president's budget has inverted the priorities with more money being spent on flagships than on any of the other categories in order to have two flagship missions to Mars in this decade, while providing funding for at most two Discovery missions in this decade. The proposed second flagship mission is also not consistent with the decadal survey's priority for the next flagship, at least to the extent that they plan a clone of Curiosity. Only a significantly modified Curiosity could support the priority of the decadal survey, namely to cache samples for return to Earth.

There is overlap between the president's proposed approach and today's issues of NEO discovery and hazard mitigation, but the administration's goals are sufficiently narrow that the overlap is not large. Finding only asteroids that would allow a short round-trip travel time to Earth is very different from finding all NEOs that might be hazardous. On the other hand, the 140-m survey under discussion here, would certainly address the needs of the human exploration program. The characterization of an NEO *via* human exploration would certainly overlap, and, if appropriately designed, could completely address, the characterization needs for mitigation for the particular type of NEO chosen for the human exploration mission. Addressing the characteristics of other types of NEOs for mitigation would not be addressed at all.

Questions from Ranking Member Eddie Bernice Johnson

1. This Committee is working on reauthorizing NASA for FY2014 and beyond. In your view, what priorities with regard to NEOs do we need to address in legislation? How do we ensure that private sector and international initiatives are effectively integrated into a global response?

In my view, the highest priority for the NEO surveys is to provide adequate funding that is not raided from the science budget (which includes most of the characterization efforts). I also note that NASA's budget is not the only source of U. S. Government funding for NEO surveys, since the LSST (Large Synoptic Survey Telescope) is being built for purely scientific reasons but with direct applicability to the NEO surveys and is funded in large part by taxpayer funds through both NSF and DOE.

I note that the surveys to 140m, aka the George E. Brown survey, search for all the NEOs larger than a given size (all is used to mean 90% as most of us understand it, since we can define 90% in a statistically reliable way but we can not define, *a priori* or even *a posteriori*, the exact number of NEOs to be found). If that given size is above the minimum size for causing substantial damage, and 140m diameter is certainly well above that minimum size, then one is not searching for the most common impactors. The number of NEOs > 140m in diameter is far smaller than the number of NEOs large enough to cause major damage but smaller than 140m. NASA has recently begun an effort to deal with the smallest ones that can cause major damage by funding a two-telescope version of the ATLAS system, which can provide enough warning of small impactors to evacuate endangered areas if they are not too large.

Noting the uncertainties that remain in mitigating the hazard from NEOs, and noting that mitigation can be seen as a threat by certain countries if undertaken by another country that is not entirely trusted, I think that the other priority in funding for an NEO program should be initiating mitigation studies. For geo-political safety, this effort should be international and should involve countries that might not trust the US fully. A basic research program on mitigation can be entirely domestic, although undertaken so as not to duplicate any non-US efforts, but practical mitigation experiments must be international. Since the key driver here is a social need, this would need to be a program funded from additional funds, not funds diverted from other NASA programs.

Integrating the private sector is straightforward for any space-based effort, since all space-based efforts discussed publicly would rely on NASA's DSN for communications. Thus NASA can easily insist on appropriate data sharing and program coordination as a condition of access to the DSN. Integrating the international effort is much more difficult. NASA officials have been good at pursuing the issues - at international scientific meetings, in discussions with other space agencies, and in activities of the UN such as COPUOS. However, ensuring international collaboration will likely require effort on the more purely political side, be it the State Department or the Office of the President or some other agency, since it requires decisions by other sovereign entities to commit substantial funds. NASA can ensure collaboration with some other space agencies, such as ESA and JAXA, in scientific space missions since those are all clearly in the mandate of those agencies and the U.S. has significant assets to trade in the DSN and in the scientific and technical work force of the U.S.

> 2. What are the risks, if any, of relying on non-government organizations to provide data needed to meet a congressional mandate? If such non-governmental capabilities are delayed or become unavailable, what options would the government have to obtain the needed data.

There are risks in any approach to meeting the congressional mandate regarding surveys for NEOs. Space missions are inherently risky but have high payoff. Ground-based telescopes are less risky and much lower cost, but also less efficient in carrying out the search. The major risk for any approach, even one funded directly by the US government, is the funding schedule. A key management lesson I learned in running the Deep Impact project is that one has to plan for the "unknown unknowns". -- the technical issues that are totally unforeseen. There are more of these in space missions than in building a ground-based telescope, but they occur in all projects involving new technology. Funding and extra schedule margin to deal with these "unknown unknowns" must either be built in at the outset or the project may run out of either funding and schedule. One of the known unknowns, i.e. a known scenario that might or might not occur, is a funding risk, whether it is withdrawal of a donor from an NGO or sequestration of government funding or inconsistent appropriations.

The risks are primarily on the construction and deployment side with smaller risks due to inadequate data pipelines. Risks of non-delivery or delayed delivery of data, once the data are taken, are quite small and, in past experience have been a problem primarily in ventures like PanSTARRS that are funded by the U. S. military. NGOs, such as the B612 Foundation and the LSST Corporation, have no motive to withhold data, nor do most international partners. They are limited first by funding risks and secondarily by technical risks.

The best strategy for ensuring against failure of an enterprise operated by an NGO (or by an international partner) is not to duplicate the effort but rather to be investing in alternative projects that would serve the NEO discovery goal, perhaps less efficiently, but could also provide other benefits, such as a scientific return or even an additional mitigation return such as in characterization of the NEOs.

> 3. What are the key challenges to meeting Congressional direction on surveying, detecting and characterizing near-Earth objects equal to or greater than 140 meters in diameter by 2020? In your opinion, is this a technological issue or are budgetary resources the pacing item?

As noted in my reply to Representative Pallazo, I am doubtful that anything can be done at this late date to complete the survey to 140 meters by 2020. If substantial funding had been provided at the time of our NRC report, it would have been possible to complete the survey not much beyond 2020, but the time to develop the hardware for surveys limits the speed with which the survey can be completed. Funding becomes the critical step in minimizing the time to completion but, absent a crash program like the Manhattan project, it is the normal engineering and construction process that limits the time scale for completion. The B612 Foundation proposed to complete the survey by the about 2023 or 2024, but that assumes that they will be able to obtain sufficient funding to match their desired spending profile. The LSST is not scheduled to even be operational until 2021. Other potential survey satellites, such as the one that has been proposed based on WISE and NEOWISE, would need years to develop and launch followed by years to conduct the survey, so such a satellite would not complete the survey until some time in the next decade.

4. What are the challenges involved in assimilating NEO detection and characterization input from multiple observing platforms? How could this be done?

The assimilation of data for the detection of NEOs is already working well since virtually all relevant observers around the world immediately report their results to the IAU's MPC. The observers themselves analyze the images to determine accurate positions and transmit those positions to the MPC, usually in near-real time and almost always within 24 hours. The staff-members at the MPC are familiar with the vagaries of the many different platforms used for discovering NEOs and of the individual observers. The MPC is in close collaboration with the NEO program office at JPL for newly discovered objects that turn out to be NEOs. The only challenge is ensuring that the MPC is adequately funded for computing power and staff – according to the director of the MPC (in a private conversation some months ago), they are adequately funded now and the currently planned budget is adequate for the foreseeable future. Their entire funding comes from NASA.

The assimilation of data for the characterization of NEOs generally is not in such good shape. The observational data, obtained primarily under scientific research programs, must be analyzed by the individual observer since the observer is the only one who knows enough details about the instrument and the observational procedures to do the analysis. These analyses, which are much more varied than the analysis of images for positions of new objects, can take anywhere from a day or two for particularly interesting cases, to years for larger surveys that assemble data on many objects before analyzing them. The results are then published in the refereed scientific literature, becoming widely available anywhere from 6 months to many years after the data were obtained. For particular cases of interest (such as the target of the OSIRIS-Rex mission), results are often made available very quickly. Some of the observers also send their data to the Small Bodies Node (SBN) of NASA's Planetary Data System, which archives all data related to small bodies, including NEOs, from NASA's missions, but which also accepts data voluntarily submitted by individual researchers. Since the U. S. Government (primarily but not exclusively through NASA) funds only a small fraction of the observers who characterize NEOs, there is financial leverage only over this small fraction to analyze the data quickly and provide public data quickly.

The SBN is the most obvious repository for most of the data and even non-US investigators provide data to SBN, but the fraction of observers who do so is small. Some of the key results, particularly the radar studies from Arecibo and Goldstone, tend to provide their data directly to the NEO program office rather than to SBN, although some aspects of the data are also submitted to SBN. Small steps could be requiring US funded investigators to archive their data publicly (at SBN) within a certain time (which might require extra funding to those observers), coupled with a special effort by SBN to reach out to the non-US observers on a regular basis (an effort which is not within the current scope of SBN). One could work with non-US space agencies to have them archive characterization data from their regions but this is entirely outside the present scope of those non-US archiving groups.

5. How did Deep Impact, the science mission for which you were the principal investigator, contribute to understanding mitigation approaches? Are there other potential opportunities for demonstrating mitigation techniques while also carrying out high-priority science investigations?

On the purely technology side, Deep Impact demonstrated the auto-navigation technology for targeting a small body of arbitrary, unknown shape and rotational state. Deep Impact also demonstrated the importance of rapid orientation corrections since tiny (a small fraction of an ounce) pieces of debris in the vicinity of the comet or asteroid can produce large pointing errors in a 1/3 ton spacecraft at the high approach speeds needed for a kinetic impactor and likely unavoidable for a nuclear blast.

On the scientific side, it is important to note that there is a wide variety of properties among the NEOs. Ideally for mitigation one would like to know the physical properties of the actual impactor, but for now we can only know the range of properties and whether those properties correlate with quantities easily measured from Earth. Results from Deep Impact are applicable only to the 10% or so of NEOs that are thought to be extinct cometary nuclei. Among the scientific results with technological implications, Deep Impact showed that cometary nuclei are extremely porous – 50% or more empty space inside the object. Deep Impact also allowed a rough estimate of the momentum transfer efficiency, roughly 2 (the range can be from 1 to about 10), a key parameter for predicting the effectiveness of kinetic impactors and one that is closely related to the properties needed to determine the effect of a nuclear blast. There is some evidence, not yet fully accepted, that the impact triggered additional spontaneous outgassing at the impact site. If confirmed, the implications for mitigation need to be explored since the active area points in different directions as the object rotates so the net effect could be to enhance the deflection or it could be to decrease the deflection.

Questions submitted by Rep. Steve Stockman

- What are the key technology demonstrations that would need to be conducted for deflecting asteroids?

Since there is more than one deflection technique, each appropriate in a different portion of parameter space (see discussion in response to the question about range). For a gravity tractor, the technology of station keeping while pulling the asteroid and while the asteroid rotates, changing the gravitational force on the spacecraft, needs to be demonstrated. For a kinetic impactor, the technology of targeting a small and/or rapidly rotating NEO needs to be demonstrated. The technology of controlling the deflection also needs to be demonstrated. For a nuclear stand-off blast, one would have to violate numerous international treaties to demonstrate the technology. Fortunately, the targeting technology is the same as for the kinetic impactor, while we would need to rely on theoretical simulations for the effect, just as we rely on such simulations for upgraded nuclear weapons.

- What is the state of the readiness of the technology for the various methods of deflecting asteroids?

Some of the technologies are well understood but others are not. Navigation to an NEO is routine in NASA's Planetary Science Division, provided the orbit of the target is well known. Auto-navigation at high speed to deliver either an impactor or a nuclear weapon was demonstrated by NASA's Deep Impact mission, but this particular target, chosen for scientific reasons rather than for mitigation testing, was a relatively "easy" target in being large and in a well defined orbit and in having a relatively simple shape. The technology for the case of an NEO with a poorly defined orbit, or a very small NEO (and small ones are the most common ones), or one with a very elongated shape (such as the nucleus of comet Hartley 2), or very fast spin period has never been demonstrated. The largest uncertainty for either a nuclear blast or a kinetic impactor is the nature of the target and the key parameters are expected to span a wide range among the NEOs.

- What would be the effective range (in time/distance) of applicability of each method of deflection?

It is important to note that the various techniques for deflection have very different upper limits, but the lower limit is similar for all the techniques. We assume that actual mitigation would use the lowest capability technique that is adequate, because this would also be the most precise and reliable technique. In other words, each technique works up to a certain size NEO and down to a certain advance warning time. The length of the warning time needed depends critically on the level of readiness to launch, so I will address only the time needed from actual launch time. I will also assume capabilities for launch vehicles, in-space propulsion, and so on that are foreseeable in the next decade independent of any NEO mitigation program. The two key factors are the time to get to the target and the time to act on the target. Ideally the deflection is applied very far from Earth since this minimizes the deflection needed - this could be on the opposite side of the sun or beyond the orbit of Mars, so that a small change in the orbit of the NEO leads to a large change in the place or time at which the NEO intersects Earth's orbit. Travel times to the target could be several months but could range up to a couple of years depending on the details of the NEO's orbit.

A nuclear standoff blast (or possibly several such blasts from a succession of launches) can work for the largest known NEOs, up to 10-20 km, with advance warning of 5 years and can work for medium NEOs, with advance warning of order a year. Kinetic impacts can work for NEOs up to about 1-2 km with several decades of warning and for 100-m NEOs with a few years of warning. Slow push/pull techniques, of which the gravity tractor seems to be the most straightforward, can work for NEOs up to a few 100 m with many decades of warning and for NEOs of 100m with 1-2 decades of warning. Many NEOs have orbits that approach Earth and then go on to hit Earth only if they pass through a relatively small region, often called a keyhole, where they are deflected by Earth's gravity such that they hit Earth years or decades later. In this case, the deflection required is small compared to what is needed for NEOs that do not have keyholes, so any technique can work with a much shorter warning time. There is a practical lower limit on the warning time for any NEO of up to a year or two because of the flight time to reach the NEO before it is too close to be deflected. The only possible exception might be an NEO small enough to be entirely destroyed into very small pieces, instead of being deflected, by a nuclear blast. The figure below, taken from the NRC report, shows the typical range of applicability in time and size (but not in distance from Earth). The boundaries are fuzzy both because of uncertainties in our knowledge and because of intrinsic variations of physical properties from one NEO to another.

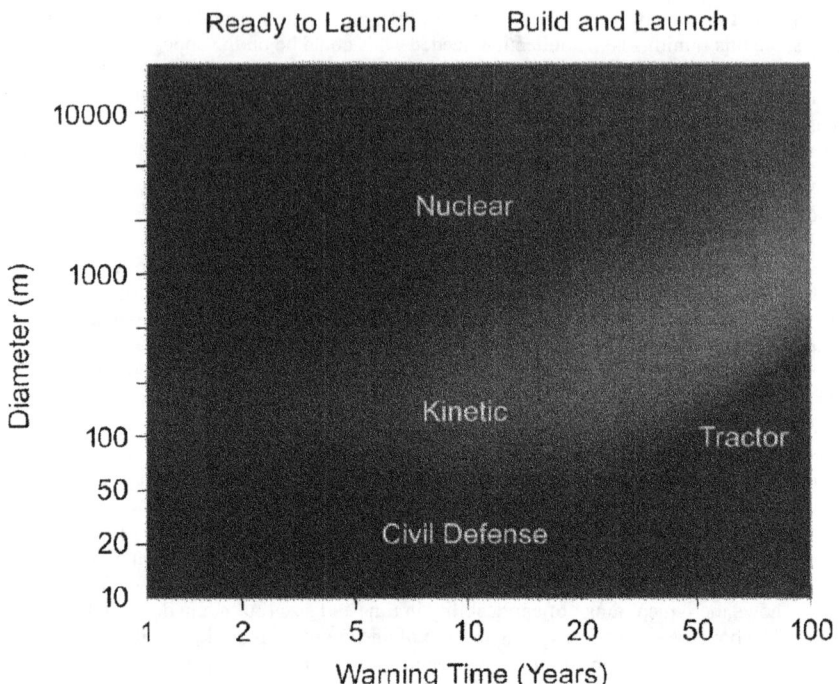

Appendix II

ADDITIONAL MATERIAL FOR THE RECORD

SUBMITTED STATEMENT BY REPRESENTATIVE STEVE STOCKMAN, COMMITTEE ON SCIENCE, SPACE AND TECHNOLOGY

Thank you, Mr. Chairman, for this opportunity to address the Subcommittee on this most important topic.

Our Nation—indeed, our world—faces threats in the future from asteroids too small to be detected by our present means but large enough to do unspeakable damage to our population centers. Witness the recent event in Russia, which has raised worldwide awareness of the potential threat. Clearly, it is important for us to increase the sophistication of our space sensors so we can detect them in advance. But once we spot them, we must ask: what can we do to protect the people?

This Committee is to be commended for its recognition of this important issue, and I appreciate this opportunity to address a unique technology that could provide us with a potential arrow for our quiver of planetary defense. That unique capability is being developed by a small, high-technology American company named Ad Astra Rocket, located in Webster, Texas next to NASA's Johnson Space Flight Center.

The company is perfecting the "VASIMR" plasma rocket engine, a game-changing electric propulsion system which originated at NASA under the leadership of its inventor, Dr. Franklin Chang Díaz, a former NASA astronaut. Incidentally, more than four years ago, during his U.S. Senate confirmation hearing testimony, NASA Administrator Charles Bolden described the 25-year efforts of Dr. Chang Díaz and his small team at NASA who kept this technology alive on only "a small stipend" from NASA.

Since it spun off from NASA in 2005, Ad Astra has continued VASIMR development—at a much faster pace and exclusively with private funds that brought the technology to a high state of maturity. At a power level of 200 kilowatts, their prototype is one of the most powerful plasma rockets operating in the world. It has been fired reliably more than 10,000 times in their vacuum chamber.

I know this is not the only advanced rocket being studied today. Other technologies, such as hall thrusters and ion engines, are being developed by NASA. However, while NASA remains an American space technology powerhouse, the world has changed since the opening of the space age in the 1950s and 60s, and U.S. innovation in rocket technology is no longer confined to NASA. It exists as well in small entrepreneurial start-ups such as Ad Astra and others that help maintain our nation's technological edge razor sharp. The government must keep pace with this changing paradigm and resist becoming a de-facto competitor with the private sector. It must ensure that fair and open competition is promoted and supported at all levels. Judging from its recent performance results, the VASIMR technology certainly deserves the opportunity to show what it can do.

Now one of those potential missions—and the major focus of our hearing today—is rocket technology to help us avoid a near Earth asteroid collision by deflecting it away from the Earth. In response to this, Ad Astra recently undertook a study on how this might be accomplished. Their concept involves a solar powered robotic craft, propelled by Ad Astra's high power VASIMR rockets that, upon arriving at the asteroid, uses the plasma exhaust of one of its two engines (the other is used to keep the craft in place) to gently push the object for days or weeks, depending on the asteroid's trajectory. A recent numerical simulation successfully demonstrated the deflection of a 40,000 ton asteroid similar to the one that barely missed Earth last February and larger than the one that actually hit Russia. In their study, the team also assumed—as it actually happened—only one year advance warning to execute the mission. This was just an initial concept evaluation. The team is now further developing the full range of their mission capability. The rocket used is the 200kW VASIMR VF–200, the same model being tested in the laboratory today, and the same model the company wishes to test on the ISS in 2016.

The technology has multiple applications which go far beyond asteroid deflection, and include more economical space station re-boost, satellite deployment, retrieval and mitigation of orbital debris. This propulsion technology also enables larger payloads and much faster robotic and ultimately human missions to Mars and other points in deep space.

The Company's next step is to test the engine on the International Space Station in early 2016—a test which will validate the technology for commercial use. Ad Astra has signed an agreement with NASA to move forward on this test. As a National Laboratory, the U.S. portion of the ISS offers a unique test environment for this technology, and beyond accomplishing this important demonstration, Ad Astra's proposed electric power and propulsion test facility would actually enhance the ISS research infrastructure by providing an unprecedented power storage capability that would enable other high power experiments of great importance to developing a robust human space exploration framework.

Ad Astra continues to commit its resources to achieving this critical milestone. In my opinion, this is a valuable technology for NASA to invest in, both for the planned 2016 validation test on ISS, as well as for asteroid deflection and space debris cleanup. With such investments, the VASIMR team is prepared to step forward and undertake a number of game-changing near-term missions for NASA and the commercial space sector. These will help maintain U.S. innovation and leadership in the new frontier of commercial space and ultimately help pave the way for a robust and economically sustainable exploration of the solar system.

At a recent hearing before this Committee on asteroids, a number of experts were concerned that there were no good answers or solutions on the horizon for dealing with the threats from asteroids. Mr. Chairman, American ingenuity, such as the VASIMR electric propulsion technology, will lead the way as part of the solution to the threat from asteroids.

Thank you, Mr. Chairman.

SUBMITTED STATEMENT BY REPRESENTATIVE DONNA F. EDWARDS

It was clear from the first hearing the Committee held on this issue a few weeks ago that the problem of near-Earth objects (NEO) impacting Earth and possibly causing great harm is worrisome but preventable—if we put our minds and resources into it.

It is also clear that this Committee has been at the forefront of ensuring that NASA be tasked with detecting such NEOs.

Unfortunately, it appears that at the present time, we still have a way to go.

Just take what recently transpired in Hawaii.

According to media reports, construction and staff jobs at the Pan-STARRS telescope system in Hawaii, which is used for near-Earth object observation, among other purposes, had to be rescued by an anonymous $3 million donation after federal funding was cut.

Imagine that, a capability critical to saving the world from potentially hazardous asteroids needed to be saved by a private donor.

But wait, it doesn't stop there. Because of the recent sequester, NASA is suspending, effective immediately, all education and public outreach activities. In terms of scope, this includes all education and public outreach efforts conducted by programs and projects.

Needless to say, it will be hard to increase public awareness of what NASA is doing in detecting NEOs under this suspension.

At this hearing, we will hear how nongovernment entities are proposing to use their own funds to save the Earth by detecting, characterizing, and perhaps even deflecting asteroids.

Some of these entities are driven by a noble cause, to save humanity, and are banking on philanthropists to finance their efforts.

Others, who are planning to mine asteroids to extract ore and minerals, see their efforts as useful for detection and characterization, since one needs to know where these asteroids are and what their composition is likely to be before a mining mission is chosen.

Now, don't get me wrong. I think it's great if the government doesn't have to foot the entire bill for proposed missions and technologies.

But what happens when something does not work, or when donations or investor contributions do not materialize? Is it prudent for the world to solely bank on the success of these nongovernment efforts? What happens when a private initiative is no longer an option? Would the government need to step in?

So there are a number of questions this Committee should be examining, and I look forward to hearing from our witnesses on their perspectives.

PLANETARY SOCIETY REPORT SUBMITTED BY REPRESENTATIVE ROHRABACHER

Preventing Asteroid Impact Written Testimony submitted to the United States Congress by The Planetary Society

April 9, 2013

Introduction

As the Chelyabinsk impact demonstrated with a relatively small example, asteroid impacts happen and they can be destructive. Another one is surely coming our way, so we must invest in what may be needed to prevent them hitting the Earth. The Planetary Society has been involved with the Near Earth Object (NEO) threat almost since The Planetary Society's founding in 1980. Here we review public interest in the topic, Planetary Society contributions, and recommendations for NASA and the support of NASA.

Public Interest in the Asteroid Threat

As the world's largest space interest group, The Planetary Society is uniquely positioned to evaluate the long-term public interest in preventing asteroid impact. Utilizing only Planetary Society member contributions, The Planetary Society has supported a variety of near Earth object (NEO) projects beginning shortly after the founding of the Society in 1980. Members and non-members have shown a consistent interest in the asteroid threat. As demonstrated by the current congressional hearings as well as press coverage, the events of Feb. 15, 2013 have escalated interest. The close flyby of Asteroid 2012 DA14 and the airburst and impact near Chelyabinsk Russia awoke many to the asteroid threat. Scientists and many citizens around the world have long recognized the threat, as evidenced by the thirst for knowledge on the subject and donations to support it by our membership over the last three decades.

The Planetary Society and NEOs

What do the discovery of 2012 DA14, and the most productive observatory for follow-on (orbit determining) NEO observations have in common? They were both made possible by Planetary Society Gene Shoemaker NEO Grants, now in their 16th year. The close fly-by Asteroid 2012 DA14 was only discovered in 2012 because of a grant to Spanish amateur observers at La Sagra Observatory for a telescope camera that would enable them to hunt for fast moving asteroids that can be missed by the large NASA funded professional surveys. Since their grant in 2010, the Spanish group has discovered more than 20 fast moving NEOs missed by the surveys. The amateur Astronomical Research Institute in Illinois obtains more follow-up observations of newly discovered and recovered NEOs than any other observatory on the planet. These types of observations are crucial because just discovering a NEO isn't enough – you have to have enough careful position observations to be able to determine its orbit and if it will hit Earth. These are just two examples of the many similar groups around the world who have been able to make their observations because of Shoemaker NEO grants. The Planetary Society has awarded 38 grants to observers in 15 countries on 5 continents. A new round of winners will be announced on April 17 at the Planetary Defense Conference.

The Planetary Society has also filled a range of other niches, from funding professional observers in the 1980's, to currently funding research in Scotland on a potential new asteroid deflection technique using spacecraft based lasers to vaporize rock on an asteroid to create jets of material that will change its speed and direction slightly, altering its orbit and preventing it from impacting the Earth. We also support a project finding more impact scars on Earth that tell tales of past impacts. Total NEO related grants from The Planetary Society have totaled nearly a half million dollars, which is quite a sum for our relatively small non-profit organization.

The Planetary Society also invests time and materials to educate the public about the NEO threat. We participate as a Non-Governmental Organization on Action Team 14 of the United Nations Near Earth Object Working Group, working the international aspects of the NEO threat.

NASA and NEOs

NASA itself is providing testimony about their NEO programs. The Planetary Society, representing the general public, encourages increased funding for NASA's NEO programs.

At current low funding levels for NEO work, NASA does an admirable job, including the funding of the most productive NEO discovery programs in the world. The Planetary Society, including its members, recommends increased support of all kinds, from hearty thanks (for potentially saving the world) to increased financial support.

The NEO threat is a global issue on several fronts. Disasters could be international in destruction and certainly would be international in disaster relief. Deflection of a dangerous asteroid would imperil other parts of the world before the target point moved off the Earth. Some of the most efficient deflection techniques use weapon type technology. Despite the international nature of the problem, NASA is uniquely poised to be the leader in all aspects of the asteroid threat.

We encourage NASA, and we recommend additional support for NASA, to expand its roles in all aspects of the asteroid threat:
- **Finding** – we can't prevent a disaster or even evacuate if we don't know what NEOs exist. This is already NASA's key role, but can be expanded, particularly to find smaller "city killer" sized asteroids, whether using ground based or space based assets
- **Tracking** – just knowing they are there isn't enough – we have to know if they are headed to Earth
- **Characterizing** – in addition to general science, we must understand the physical characteristics of potentially dangerous asteroids, e.g., solid or fluffy, one asteroid or a binary pair.
- **Deflection techniques** – Though finding and tracking are the first steps, right now, we have no mature deflection techniques. Asteroids are spinning or tumbling, which may render velocity changes quite difficult to achieve. To be prepared, we should develop the techniques and test them before we need them. Some techniques are better suited to larger asteroids, some to smaller, etc. We need to make sure we can make them work.
- **Political planning and agreement** – We strongly recommend establishing international agreements now. Beginning negotiations after we find an asteroid headed toward Earth could prove disastrous. These activities have at least now started, i.e., through the United

Nations, but there is much more that can be done, including the various processes that must be put in place depending on the composition of the object, its spin, its size, and the length of our warning time,

Conclusions

The Planetary Society will continue to find niches where we can contribute in the world of NEOs. Our nature as a public-supported, international group allows us nimbleness and flexibility, e.g., to support qualified amateurs around the world, or to provide quick, critical seed funding to jump-start an interesting project that needs a boost. But, there are many things that only a national space agency can address, whether due to financial scale, or due to the government level representation with the international community.

Asteroid impact stands as the only preventable natural disaster. In the wake of the events of February 15, the time is ripe for action. We must invest in asteroid detection and mitigation for the sake of all humankind.

Æ